AF878972

Application of Calorimetry in Life Sciences

Application of Calorimetry in Life Sciences

Proceedings of the International Conference
in Berlin, August 2-3, 1976

Editors
I. Lamprecht
B. Schaarschmidt

W
DE
G

Walter de Gruyter · Berlin · New York 1977

Editors
Dr. Ingolf Lamprecht
Dr. Bernd Schaarschmidt
Zentralinstitut für Biochemie und Biophysik
Freie Universität Berlin
Habelschwerdter Allee 30
D-1000 Berlin 33

with 243 Figures

CIP-Kurztitelaufnahme der Deutschen Bibliothek

Application of calorimetry in life sciences:
proceedings of the internat. conference in Berlin,
August 2–3, 1976/ed.: J. Lamprecht; B. Schaarschmidt. –
1. Aufl. – Berlin, New York: de Gruyter, 1977
ISBN 3-11-006919-9
NE: Lamprecht, Ingolf [Hrsg.]

© Copyright 1977 by Walter de Gruyter & Co., Berlin 30.

All rights reserved, including those of translation into foreign languages. No part of this book may be reproduced in any form by photoprint, microfilm, or any other means, nor transmitted nor translated into a machine language without written permission from the publisher. Printing: Karl Gerike, Berlin. Binding: Lüderitz & Bauer, Buchgewerbe GmbH, Berlin. Printed in Germany.

Preface

All methods of modern physics and chemistry are now applied to biology and will be increasingly valuable for biological research in the near future, but no other field has such ever increasing importance and promise for a deeper understanding of life and its traces as thermodynamics, and especially the theory of irreversible processes. One essential tool of thermodynamics is calorimetry, calorimetry in its application to life.

Unlike optical, enzymatic, isotopic or ultrasonic tests, calorimetry does not interfere directly with the object under research, since nearly all chemical and biochemical processes are combined with a production or consumption of heat and therefore with a flow of heat between the system and its surroundings. Modern calorimeters collect these weak flows and transform them into usable signals. It is left to the skill of the research worker to translate, understand and interpret these signals and to enlarge the information about the system. Unfortunately, however, heat is dreadfully unspecific, and one needs the combination with other methods to gain a complete understanding of the processes under study.

This book is about the applications of calorimetry in life sciences. Its approach, which is nearly two hundred years old and yet very modern, is applicable to all levels of life, from the pure biochemical reaction to whole organisms or even ecosystems. The book discusses instrumentation as well as results, correlations with other techniques as well as the limits of the calorimetric method. It is intended to give a review of recent biological calorimetry to the research workers in this field and to direct the newcomer to a method which may be helpful in the solution of his own scientific problems.

The articles in this book are the content of an international conference on the "Application of calorimetry to life sciences" which took place at the Free University of Berlin in August 1976. It was sponsored by the Zentralinstitut für Biochemie und Biophysik and financially supported by the President of the Free University and some companies, whose assistance we gratefully acknowledge. But this conference would have been impossible without the steady help and interest of all the colleagues of our institute to whom we are deeply indebted.

To facilitate and expedite the appearance of the book, the authors were asked for camera ready manuscripts. Therefore, only small corrections could be made by the editors, a disadvantage that is more than compensated, in our opinion, by the directness and personal style of the papers. We want to express our thanks to all authors for their care in the preparation of the manuscripts and their efforts to meet the deadlines imposed.

We do hope that this book will find an interested audience and help to promote our knowledge of life in its abundance of forms.

Berlin, January 1977

Ingolf Lamprecht
Bernd Schaarschmidt

Contents

List of Participants

Contributors' names are printed in bold face.

Anders, A.,	Fakultät für Physik, Universität Bielefeld, Universitätsstraße 1, D-4800 Bielefeld
Baisch, F.,	Zentralinstitut für Biochemie und Biophysik, Freie Universität Berlin, Habelschwerdter Allee 30, D-1000 Berlin 33
Beezer, A.,	Department of Chemistry, Chelsea College, University of London, Manresa Road, London SW3 LX, Great Britain
Belaich, J.P.	Laboratoire de Clinic Bactérienne, C.N.R.S., F-13247 Marseille Cedex 2
Benzing, H.,	Physiologisches Institut, Universität Tübingen, Gmelinstraße 5, D-7400 Tübingen
Berthe-Corthi, L.,	Fachbereiche Naturwissenschaften, Universität Oldenburg, Ammerländer Heerstraße 67–99, D-2900 Oldenburg
Beyersbergen van Henegouwen, H.,	LKB-Produkten BV, Zeekant 35, NL-Den Haag
Blanc, B.,	Eidgenössische Forschungsanstalt für Milchwirtschaft, Sektion Physik und Biophysik, CH-3074 Liebefeld-Bern
Boivinet, P.,	Institut de Physico-Chimie Biologique, Université de Caen, F-14032 Caen
Brettel, R.,	Zentralinstitut für Biochemie und Biophysik, Freie Universität Berlin, Habelschwerdter Allee 30, D-1000 Berlin 33
Bubenger, H.J.,	Institut für Diabetesforschung, Auf'm Hennekamp 65, D-4000 Düsseldorf
Buschmann, H.	Strahlenbiologisches Institut, Universität München, Schillerstraße 42, D-8000 München 2
Crueger, W.,	Abteilung Verfahrensentwicklung Biochemie der Bayer AG, Friedrich-Ebert-Straße 217, D-5000 Wuppertal 1
Farhangi, Y.,	Chemistry Department, Faculty of Science, University of Azarabadeghan, Tabriz, Iran
Ferroni, G.,	Université de Provence, Centre Saint Charles, F-13274 Marseille
Forrest, W.,	The Australian Wine Research Institute, Private Bag, Glen Osmond, S.A. 5064, Australia
Fujita, T.,	Chemical Center, Thermochemistry, Lund University, P.O.B. 740, S-220 07 Lund

Galea, J.,	Université de Provence, Centre Saint Charles, F-13274 Marseille
Gazith, J.	Haut- und Poliklinik der Freien Universität Berlin im Rudolf-Virchow-Krankenhaus, Augustenburger Platz 1, D-1000 Berlin 65
Gnaiger, E.,	Institut für Zoophysiologie, Universität Innsbruck, Peter-Mayr-Straße 1a, A-6020 Innsbruck
Gravelle, P.C.,	Institut de Recherches sur la Catalyse, CNRS, 39 boulevard du 11 novembre 1918, F-69626 Villeurbanne
Gruber, K.,	Institut für Zoophysiologie, Universität Innsbruck, Peter-Mayr-Straße 1a, A-6020 Innsbruck
Gustafsson, L.,	Botanical Institute, Department of Marine Microbiology, Carl Skottsbergsgatan 22, S-41319 Göteborg
Held, W.,	Lehrstuhl für Biotechnologie, Technische Universität Berlin, Seestraße 13, D-1000 Berlin 65
Hellwig, G.,	Institut für Grenzflächen- und Bioverfahrenstechnik, Eierstraße 46, D-7000 Stuttgart
Höpcke, R.,	Institut für Forschung und Entwicklung der Maizena GmbH, Knorrstraße 1, D-7100 Heilbronn a.N.
Jannsen, L.H.M.,	Pharmaceutical Laboratory, School of Pharmacy, Catharijnesingel 60, NL-Utrecht
Jarrett, I.G.,	Division of Human Nutrition, C.S.I.R.O., Kintore Avenue, Adelaide, S. A. 5000, Australia
Jéquier, E.,	Départment de Médicine, Centre Hospitalier Universitaire Vaudois, CH-1011 Lausanne
Joly, R.,	Chemistry Department, I.N.S.A., 20 Avenue Albert Einstein, F-69621 Villeurbanne
Kafka, H.,	LKB Instrument GmbH, Odoerkergasse 25, A-1160 Wien
Korver, O.,	Department of Spectrometry, Unilever Research, P.O.B. 114, NL-Vlaardingen
Körber, F.,	Zentralinstitut für Biochemie und Biophysik, Freie Universiät Berlin, Arnimallee 22, D-1000 Berlin 33
Krisam, G.,	Lehrstuhl für Allgemeine Chemie und Biochemie, Technische Universität München, D-8050 Freising-Weihenstephan
Lamprecht, I.,	Zentralinstitut für Biochemie und Biophysik, Freie Universität Berlin, Habelschwerdter Allee 30, D-1000 Berlin 33

Laskowski, W.,	Zentralinstitut für Biochemie und Biophysik, Freie Universität Berlin, Habelschwerdter Allee 30, D-1000 Berlin 33
Leiseifer, H.P.,	Kernforschungsanlage Jülich, Institut für Biophysikalische Chemie, ICH 2 Postfach 1913, D-5170 Jülich
Lieth, H.,	Department of Botany, University of North Carolina, Chapel Hill, N.C. 27514, USA
Lindman, B.,	Botanical Institute, Department of Marine Microbiology, Carl Skottsbergsgatan 22, S-41319 Göteborg
Lukesch, A.,	Eidgenössische Forschungsanstalt für Milchwirtschaft, Sektion Physik und Biophysik, CH-3074 Liebefeld-Bern
Monk, P.,	Chemical Center, Thermochemistry, Lund University, P.O.B. 740, S-220 07 Lund
Moor, U.,	Eidgenössische Forschungsanstalt für Milchwirtschaft, Sektion Physik und Biophysik, CH-3074 Liebefeld-Bern
Moser, H.,	Institut für Zoophysiologie, Universität Innsbruck, Peter-Mayr-Straße 1a, A-6020 Innsbruck
Müller-Schauenburg, W.,	Nuklearmedizinisches Department im Medizinischen Strahleninstitut, Universität Tübingen, Im Röntgenweg 11, D-7400 Tübingen
Mukherjee, B.,	Abteilung Nuklearmedizin im Klinikum Steglitz, Freie Universität Berlin, Hindenburgdamm 30, D-1000 Berlin 45
Nunomura, K.,	The Institute of Applied Microbiology, University of Tokyo
d'Oleire-Oltmanns, W.,	II. Zoologisches Institut, Universität Erlangen-Nürnberg, Bismarckstraße 10, D-8520 Erlangen
Ortner, B.,	Institut für Zoophysiologie, Universität Innsbruck, Peter-Mayr-Straße 1a, A-6020 Innsbruck
Oster, O.,	Zentralinstitut für Biochemie und Biophysik, Freie Universität Berlin, Ostpreußendamm 111, D-1000 Berlin 45
Perrin, J.H.,	Pharmaceutical Laboratory, School of Pharmacy, Catharijnesingel 60, NL-Utrecht
Probst, L.,	Landesanstalt für Umweltschutz Baden Württemberg, Institut für Seenforschung und Fischereiwesen, Untere Seestraße 81, D-7994 Langenargen
Ragot, M.,	Laboratoire Chimie Bacterienne, C.N.R.S., 31 Chemin Joseph-Auguier, F-13274 Marseille
Reddy, A.R.,	Abteilung Nuklearmedizin im Klinikum Steglitz, Freie Universität Berlin, Hindenburgdamm 30, D-1000 Berlin 45

Ric, S., Chemical Physics Department, Weizmann Institute of Science, Rehovot, Israel

Rohdewald, P., Institut für Pharmazeutische Chemie, Universität Münster, Hittorfstraße 58–62, D-4400 Münster

Rüegg, M., Eidgenössische Forschungsanstalt für Milchwirtschaft, Sektion Physik und Biophysik, CH-3097 Liebefeld-Bern

Samuelsson, E.G., Dairy Department, Royal Veterinary and Agriculture University, Kopenhagen, Denmark

Sari, J.C., Laboratoire Chimie Bacterienne, C.N.R.S., 31 Chemin Joseph-Auguier, F-13274 Marseille

Sayyadi, P., Zentralinstitut für Biochemie und Biophysik, Freie Universität Berlin, Habelschwerdter Allee 30, D-1000 Berlin 33

Seidel, N., Soldiner Straße 37, D-1000 Berlin 65

Semeritz, E., Bundesst. Bakt.-Serol.-Untersuchungsanstalt und Institut für Hygiene Universität Innsbruck, A-6010 Innsbruck

Slotboom, A.J., Laboratory of Biochemistry, University of Utrecht, Transitorium III, Padualaan 8, NL-Utrecht

Suurkuusk, J., Chemical Center, Thermochemistry, Lund University, P.O.B. 740, S-220 07 Lund

Schaarschmidt, B., Zentralinstitut für Biochemie und Biophysik, Freie Universität Berlin, Habelschwerdter Allee 30, D-1000 Berlin 33

Schachinger, L., Abteilung Strahlenbiologie und Biophysik, Institut für Biologie, Ingolstädter Landstraße 1, D-8042 Neuherberg

Schaefer, H., Haut- und Poliklinik der Freien Universität Berlin im Rudolf-Virchow-Krankenhaus, Augustenburger Platz 1, D-1000 Berlin 65

Schildknecht, J., Hoffmann-La Roche und Co. AG, CH-5058 Basel

Schmidt, H.-L. Lehrstuhl für Allgemeine Chemie und Biochemie, Technische Universität München, D-8050 Freising-Weihenstephan

Schnabel, C., LKB Instrument GmbH, Lochhamer Schlag 5, D-8032 Gräfelfing

Stein, W., Rheinbabenallee 3, D-1000 Berlin 33

Taylor, R., LKB Instruments Ltd., 232 Addington Road, South Croydon, Surrey CR 2 84D, Great Britain

Tiefenbrunner, F., Institut für Hygiene und Mikrobiologie, Universität Innsbruck, Fritz-Pregel Straße 3, A-6010 Innsbruck

Trumpa, M.,	Zentralinstitut für Biochemie und Biophysik, Freie Universität Berlin, Habelschwerdter Allee 30, D-1000 Berlin 33
Tumerman, L.,	Department of Chemical Physics, The Weizmann Institute of Science, P.O.B. 26, Rehovot, Israel
Vonier, M.,	Messgeräte Vertrieb, St.-Martin-Straße 30, D-8061 Kleinberghofen bei München
Wadsö, I.,	Chemical Center, Thermochemistry, Lund University, P.O.B. 740, S-220 07 Lund
Welge, G.,	Zentralinstitut für Biochemie und Biophysik, Freie Universität Berlin, Habelschwerdter Allee 30, D-1000 Berlin 33
Wendt, B.,	Zentralinstitut für Biochemie und Biophysik, Freie Universität Berlin, Habelschwerdter Allee 30, D-1000 Berlin 33
Weßelmann, G.,	Institut für Pharmazeutische Chemie, Universität Münster, Hittorfstraße 58–62, D-4400 Münster
Woledge, R.C.,	Department of Physiology, University College London, Gower Street, London WC1, Great Britain
Yasui, A.,	Zentralinstitut für Biochemie und Biophysik, Freie Universität Berlin, Habelschwerdter Allee 30, D-100 Berlin 33
Zidovezki, R.,	Chemical Physics Department, Weizmann Institute of Science, Rehovot, Israel
Zotin, A.I.,	Institute of Developmental Biology. Academy of Science of the USSR, 26 Vavilov St., Moscow 11 7334

1. Physical Calorimetry and Instrumentation

1.1 The Microcalorimetric Study of Catalytic Reactions

P. Gravelle

When I received Professor LAMPRECHT's invitation to give the opening lecture of this Conference on the Application of Calorimetry in Life Sciences, I felt very much honoured and, at the same time, slightly anxious : my expertise in Life Sciences is limited to the pleasant memory of practical jokes played upon Professor Calvet by some of his students, placing flies or other small insects in the calorimeters and letting Pr CALVET discover the surprising calorimetric curves thus obtained!

This talk will therefore be limited to a review of the uses of microcalorimetry in the study of catalytic reactions. Although this topic may appear very foreign from the interests of many participants to the Conference, there is, at least, one common feature in the use of calorimetry in heterogeneous Catalysis and in some studies in Life Sciences. In both cases, a flow or doses of a reactant (or reactants) must contact a sample of some material, in order to produce or maintain the process under study. The reacting system placed in a suitable instrument where its evolution is followed, must be connected to a gas or fluid-handling system where the reactants are properly prepared, mixed, analyzed and dosed. This requirement is particularly unconvenient when calorimetry is used to detect changes of enthalpy in the reacting system since the line connecting the calorimeter to the reactant-handling system may be the source of heat leakage, which perturbs the thermochemical measurements. Catalytic chemists have however demonstrated their skill and ingenuity in solving this and other problems concerning the use of calorimetry in this field. They have used and developed instruments based on different principles (adiabatic, isothermal, isoperibol calorimeters). Moreover since catalysis and adsorption phenomena are not limited to temperatures close to normal temperature but require excursions down to 77 K or up to 1000 K, they have had to study, with a particular care, the different experimental limitations of their instruments. Calorimetric research in Life Sciences may benefit from their experience. Calorimetric techniques used in Adsorption and Heterogeneous Catalysis will be therefore briefly presented and the advantages and limitations of some actual calorimeters will be discussed.

Any calorimetric device may be considered as composed of a cell in which the phenomenon under study takes place and of a surrounding medium (a block, a shield, or surrounding envelopes). Let assume, as a first approximation, that, during the operation of the calorimeter, i) the temperature of the surrounding medium is constant and uniform and ii) that the temperature of the inner cell is uniform. Then, the heat balance when heat dQ is produced in time dt is simply given by TIAN equation (1)

$$\frac{dQ}{dt} = p\,\theta + \mu\,\frac{d\theta}{dt}$$

where p is the heat transfer coefficient between the cell and the surrounding medium, μ is the heat capacity of the calorimeter vessel and of its contents, θ is the temperature difference between cell and surrounding medium and $d\theta$ is the temperature change caused by the production or absorption of heat, dQ. It is clear, from TIAN equation, that the total amount of heat produced or absorbed in the cell may be deduced from the area under the calorimetric curve provided that p and μ are constant during the course of an experiment and during calibration tests and actual experiments.

TIAN equation may be written, in a more general form, as :

$$f(t) = p\left[g(t) + \tau\,\frac{dg(t)}{dt}\right]$$

where f(t) is the rate law for the process under study; g(t) is the recorded calorimetric curve and τ is the time constant of the instrument. It appears that the recorded calorimetric curve is not generally a correct representation of the rate law. A distorsion occurs which is caused by the thermal lag in the calorimeter.

In order to minimize heat leakage from the calorimeter cell, it is possible to regulate the temperature of the surrounding envelope so that it remains at any time identical to that of the calorimeter proper. Following the modern trend, we shall call "adiabatic calorimeters" these instruments. In the case of a perfectly adiabatic calorimeter, the heat transfer coefficient equals zero and the total heat produced from time 0 to t is simply deduced, as shown from integrated TIAN equation, from the temperature increase in the calorimeter cell from time zero to t, μ is determined by calibration experiments.

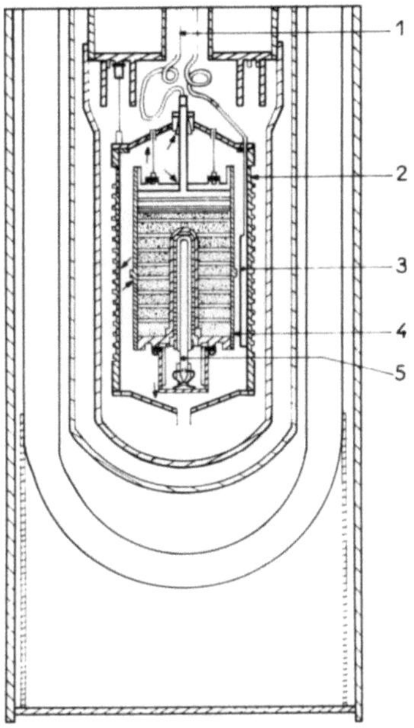

Figure 1

Adiabatic calorimeter (MORRISON and LOS, 1950) (2) 1 : filling tube; 2, adiabatic shield; 3, vapour pressure thermometer bulb; 4, calorimeter vessel; 5, platinum resistance thermometer; differential thermocouple junctions are indicated by arrows.

One of the most sophisticated calorimeters of the adiabatic type, used in adsorption studies, has been described by MORRISON and LOS (2) (figure 1). The calorimeter proper, the envelopes and attachment are placed in a liquid nitrogen or liquid hydrogen cryostat. The adsorbent is packed in the aluminum calorimetric vessel between perforated aluminum disks which facilitate heat transfer. The temperature of the calorimeter is measured by a platinum-resistance thermometer. An electrical heater wrapped around the thermometer is used for calibration. The adsorbing gas is admitted to the calorimeter via a metal capillary tube. All electrical leads and capillary tube make thermal contact with the cryostat. Heat is applied via electrical heaters on the shield (top, bottom and side), on the capillaries and bundle of electrical leads. The energy supplied to the shield is adjusted so that the differential thermocouples (marked as arrows on figure 1) give, at any instant, readings as close as possible to zero. In the absence of a thermal effect, the temperature of the calorimeter may be held constant within 0.001 K. The precision in the determination of heats of adsorption is better than 0.5 %.

In order to be measured with precision, usual values for the temperature increase during the adsorption of a dose lie within 0.5 to 2 K. Large doses of adsorbate and, consequently, large samples of adsorbent are needed. Heat and mass transfers within the adsorbent mass may be the source of errors. This difficulty, however, is not specific of adiabatic calorimeters.

The accurate determination of the temperature in the calorimeter cell, at any instant during the course of the experiment, is particularly important in adiabatic calorimetry since the amount of heat is deduced from this measurement. Resistance thermometers should apparently be preferred to thermocouples for this purpose since they provide more reliable informations about the average temperature of the cell than small-area thermocouple junctions, with an equal or better precision. Their use, however, is unfrequent because thermometers of this type continuously generate Joule heat which elevates the temperature of the calorimeter cell. This drawback has been satisfactorily minimized in the case of the adiabatic calorimeter with constant heat exchange described by KISELEV and coll. (3). In this calorimeter, the platinum resistance thermometer which is used to measure the cell temperature is wound, together with a calibration heater, around a copper envelope containing the sample in a glass vessel, connected to an external line. A second copper envelope closely surrounds the first one. Around it, another resistance thermometer is coiled, which forms one arm of a Wheatstone bridge. The second arm of the controlling bridge is a similar thermometer placed, together with a heater, on the adiabatic shield. The temperature of the shield is regulated in such a way that a constant thermal head exists, at all times, between the shield and the calorimeter proper and that the resulting heat leakage exactly balances the thermal output of the resistance thermometer. All heat generated by the process under study is however confined to the calorimeter cell and the calorimeter behaves, with respect to the phenomenon under study, as an adiabatic calorimeter of the usual type. When the calorimeter is placed in an excellent thermostat, regulated to $\pm$ 0.02 K, its temperature sensitivity is 2×10^{-5}K and its thermal sensitivity is close to 2×10^{-3} cal (4). It has been used at room temperature.

A good precision can therefore be achieved with adiabatic calorimeters. However, the use of these instruments is generally a delicate operation which requires skilled experimentators although temperature-control systems of excellent quality are now available.

Three advantages of calorimeters of this type must be cited :
a) they allow the study of slow processes. The KISELEV's calorimeter, for instance, is suitable for measurements of the heat produced in several tens of minutes;
b) the temperature of the adiabatic shield may be varied almost at will above the thermostat temperature;
c) they allow the measurement of heat capacities : heat capacities of adsorbed layers have been, for instance, determined.

Some limitations of this technique, however, should not be ignored :
a) the measurements are performed in non-isothermal conditions
b) the sample must be connected to the volumetric line via a capillary tube, in order to minimize heat leakage. The initial outgassing or pretreatment of the large samples which are used in these experiments is not easily performed and the reproducibility of this operation is doubtful.

Instead of confining to the calorimeter proper, all heat generated during the experiment, it is possible conversely to transfer as rapidly as possible the generated heat so that the calorimeter vessel remains at all times at a constant temperature. This cannot be completely true however, since a thermal head must exist for heat to be transferred but it is possible to minimize it. A phase-change in the surrounding medium is generally used to absorb the transferred heat and to determine its quantity. Simple phase-change calorimeters have been used in the past for studying adsorption but they appear to be nowadays only of historical interest.

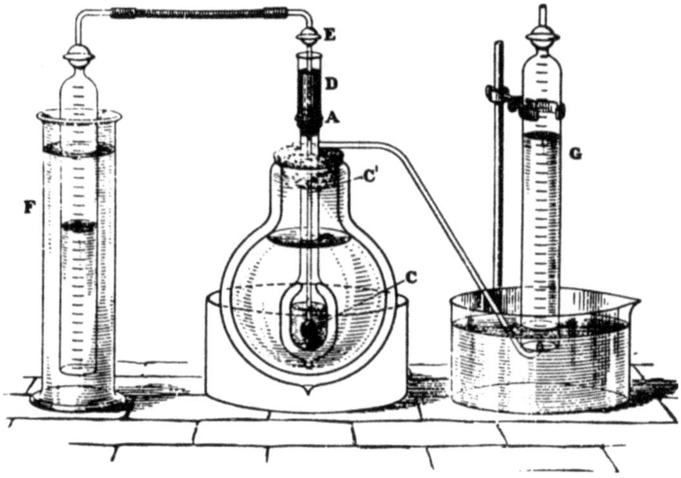

Figure 2

Phase-change calorimeter (DEWAR, 1904) (5). A, stopper; C, calorimetric liquid (liquid air); C', tube for gas entry and glass bulb containing the adsorbent; D, mercury guard; E, stopcock; F, adsorbate-dispensing burette; G, burette for measuring the volume of vaporized air.

DEWAR, for instance, used a liquid-gas transformation to study in 1904, adsorptions on charcoal at liquid air temperature (5) (figure 2). The heat evolved in the adsorption process serves to vaporize liquid air, the volume of which is measured. The heat of adsorption is therefore deduced from two volumetric measurements, that of the adsorbing gas and that of vaporized air. Both volumes being similar, they can be measured with the same type of volumetric system. Numerous difficulties due to the irregular vaporization of the liquid or small variations of its temperature limitate however the use of this technique.

Solid-liquid phase transformations have also been used for adsorption studies in BUNSEN type calorimeters. The precision of these calorimeters and their sensitivity (2 x 10^{-2} cal/mg of mercury) is acceptable but they are not very convenient to use and they operate necessarily at a limited number of fixed temperatures.

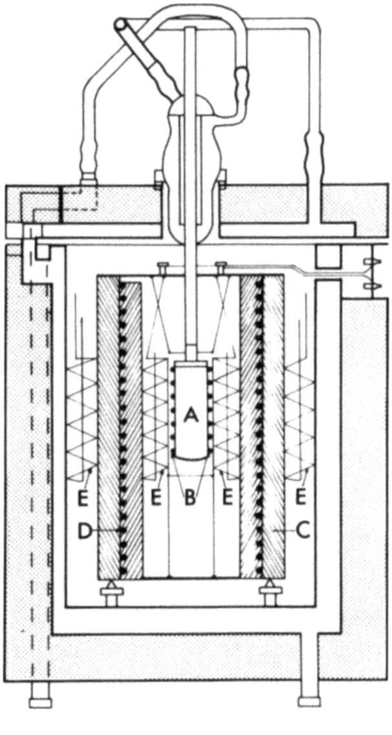

Figure 3

Isothermal calorimeter for studying the adsorption of a continuous flow of adsorbate (KISELEV et al, 1962) (6). A, calorimeter cell; B, heater; C, isothermal shield; D, heater; E, thermocouple batteries.

A more reliable and precise method for maintaining isothermal conditions is to regulate the temperature of the calorimeter vessel in such a way that it remains constant throughout the experiment. KISELEV and coll. have operated one of their adsorption calorimeters according to this principle (6) (figure 3). The inner cylindrical vessel, which contains the adsorbent, is continuously heated by passing pulses of current in a heater in such a way that a constant heat exchange is maintained at any time between the inner vessel and the outer isothermal jacket. When an exothermic reaction occurs in the calorimeter vessel, the number of pulses per unit time decreases; conversely, it is increased when an endothermic phenomenon takes place in the cell. From a record of the number of pulses during the experiment, the quantity of heat produced or absorbed by the reaction may be calculated. When the pulse generator is controlled by a quartz crystal frequency stabilizer, the temperature of the inner vessel is constant to within 10^{-5} K during a period of 6 to 8 hrs. The precision of the measurements may then be up to 0.2 %. In order to extend the temperature range of this calorimeter up to 670 K, a differential arrangement has also been utilized.

The principle of this isothermal calorimeter is simple and sound, the only source of variable heat leaks being via the gas phase in the tube which connects the sample cell to the gas handling system. In principle, there is no need to calibrate the calorimeter since this is continuously done during any experiment. Thermal paths within the calorimeter cell being short, the response of the instrument is rapid. Because of the discontinuous heating of the inner cell, this calorimeter however is not very suitable for following kinetics of thermal processes, but it may be used indifferently to study fast or slow adsorption processes at the surface of good or poor heat-conducting solids. It requires, however, high quality electronics for the temperature controls and the regulation of the power supply.

Most experimentators working in laboratories which do not specialize in Thermodynamics are not, however, prepared to build such elaborate equipments, especially when calorimetric data are needed but occasionally to confirm or complete results obtained by more conventional techniques in Adsorption or Catalysis. This is probably why most calorimeters which have been used in these fields are not adiabatic nor isothermal.

Many of these calorimeters consist of an inner vessel which is imperfectly insulated from its surroundings, usually maintained at a constant temperature. Such calorimeters are called "isoperibol calorimeters" (constant surroundings in greek). As soon as reaction takes place, a temperature gradient exists between the inner vessel and the surrounding shield and heat losses occur producing the cooling of the calorimeter vessel. When all heat is evolved instantaneously at the initial time of the experiment, it follows from NEWTON's cooling law, that the semi-logarithmic plot of the calorimeter temperature changes versus time should yield a straight line. The maximum temperature increase in the calorimeter vessel and thence the heat evolved are then obtained by extrapolating the straight line at time zero. This simple method has been used to measure heats of, fast, adsorptions. It is also possible, and more accurate, to construct the adiabatic curve for any given experiment, i.e. to determine the course of the calorimeter temperature if the instrument had been perfectly adiabatic at all times.

Calorimeters of this type have been constructed for very different studies in adsorption and Catalysis but the isoperibol calorimeter, first proposed by BEECK in 1945 (7) is one of the most thoroughly studied adsorption calorimeters of the isoperibol type (8). It is used to determine heats of adsorption at the surface of metal films.

The cell of a film calorimeter generally consists of a thin-walled tube closed at one end (figure 4). It is mounted in a glass jacket which can be maintained under reduced pressure, in order to reduce thermal leakage by conduction or convection. The jacket is placed in a thermostat. The film of the desired metal is deposited on the inner surface of the calorimeter wall by heating a filament of the metal. The temperature increase which appears as a dose of gas adsorbs on the film is generally measured, as a function of time, by means of a platinum resistance thermometer regularly wound around the calorimeter cell, in order to average all local fluctuations of temperature.

The electrical calibration of film calorimeters is achieved by passing a current of known intensity in a heater also wound around the calorimeter cell or, what apparently seems preferable, in the metal film itself. Until now, film calorimeters have been exclusively used to study adsorption processes characterized by very low equilibrium pressures. No heat is thus evacuated by the gas phase in the calorimeter and it has been shown that the main source of heat transfer in film calorimeters of this type is through radiation across the evacuated space between the cell and the jacket.

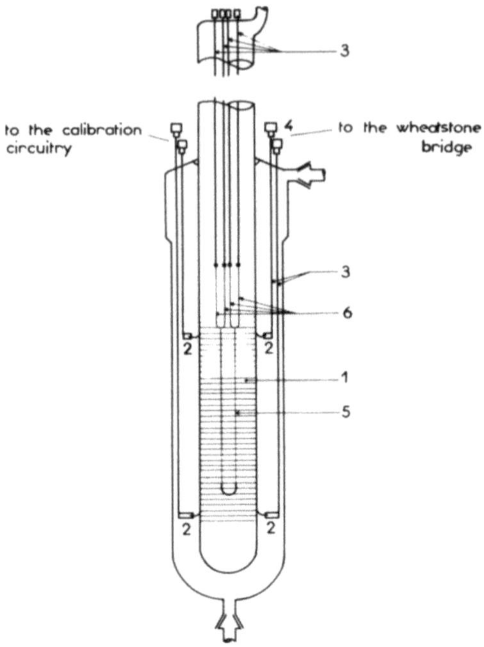

Figure 4

Isoperibol calorimeter (CERNY, PONEC and HLADEK, 1966)(9). 1, thermometric wire interwound with a calibration wire; 2, nickel foils to which the platinum thermometer wires or calibration wires respectively are spot-welded; 3, glassed molybdenum rods; 4, nickel foils, spot-welded on one end to molybdenum rods and on the other end to gilded platinum foils to which in turn the respective electrical circuitry is soldered; 5, evaporation filament; 6, thick nickel leads.

One of the drawbacks of cylindrical film calorimeters is that the admitted gas may adsorb preferentially in the upper part of the metal film and that, thereby, the adsorbed species may not be distributed uniformly on all equivalent surface sites (8). Differential heats of adsorption, thus obtained, would thus be questionable. Spherical adsorption calorimeters have been used to minimize this drawback.

A very large number of isoperibol calorimeters have also been constructed to study adsorption on powdered solids (10). In all cases, the purpose of the arrangement is clear : the calorimeter proper is thermally insulated from its surroundings in order to minimize heat leakage and thereby to maintain the temperature increase in the cell to a maximum value. A good sensitivity may then be achieved even when simple thermometers are used. But, heat leakage inevitably exists and the heat leaks in these calorimeters may follow unpredicted paths. The heat transfer coefficient for the different paths may be moreover difficult to measure and may vary in the course of the experiment. For these reasons, these calorimeters are generally used in quasi-adiabatic conditions and their use is limited therefore to the study of fast processes.

But many sensitive temperature sensors are nowadays commercially available. It does not seem necessary any longer to refrain heat leakage in order to achieve a good sensitivity. It is possible, conversely, to construct calorimeters in which a path is defined mechanically in the calorimeter arrangement for transferring heat from the calorimeter vessel to the surrounding heat sink. The heat conductor thus placed in the calorimeter facilitates the heat transfer and thence all other sources of heat leaks often contribute in a negligible manner to the total heat transfer. In this way, the heat transfer coefficient may be maintained at an almost constant value throughout the experiments and TIAN equation can be used to analyze the data provided by fast or slow processes. Calorimeters of this type are usually called conduction or heat-flow calorimeters.

The "calorimeter element" of a typical heat-flow calorimeter, the TIAN-CALVET microcalorimeter is presented on the figure 5 (1). It is composed of a cell, of surrounding batteries of thermocouples and of a heat sink. In order to increase the stability of the thermocouple emf, two such elements are usually placed in the same metal block and connected differentially. The stability of the record base-line is thus extremely good and fast or very slow phenomena can be studied. Many different versions of this calorimeter have been proposed for use at temperatures extending from

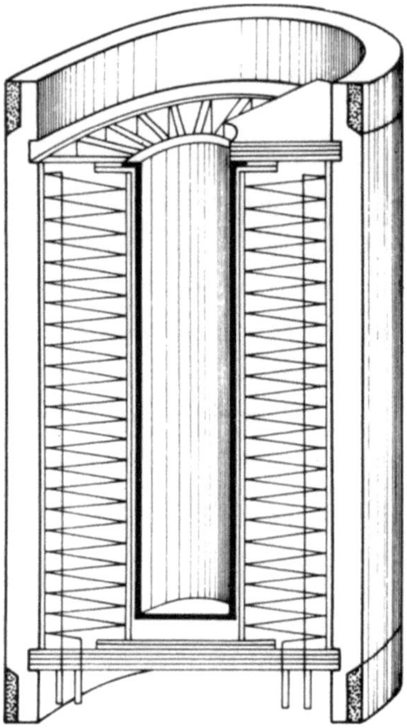

Figure 5

Heat-flow calorimeter : calorimeter element of TIAN-CALVET type, 1956(1).

77 K to 1200 K and a given calorimeter may frequently be utilized in a broad temperature range. The main disadvantage of heat-flow calorimeters is, however, that their response is frequently slow. This is caused by the large heat capacity of the calorimeter proper and temperature detectors. The profile of the thermal curves is therefore distorded by thermal lags in the calorimeter and quasi-simultaneous heat phenomena though kinetically separate may be totally blurred on the recorded thermogram.

It is clear from the preceding review that experimentators working in the fields of adsorption and catalysis have constructed and used very different types of calorimeters. It is possible to determine, from these numerous studies, the advantages and limitations of the different calorimetric techniques, as we have attempted to do. Most of these conclusions are probably not only valid in the field of Adsorption and Catalysis but should also apply to studies in various branches of the Life Sciences.

Isothermal calorimeters with power compensation and heat-flow calorimeters present a number of advantages for the study of catalytic reactions and few limitations. They should be preferred, whenever it is possible, to calorimeters of other types. In order to more specifically illustrate this point, a description of several applications of heat-flow calorimetry in the study of catalytic reactions is presented hereafter.

The heat produced by a catalyzed reaction, once it has reached a stationary state, is evidently equal, in absolute value, to the change of enthalpy which can be calculated, from Thermodynamics, for the same reaction in the homogeneous phase at the same temperature. Experiences show, however, that in most cases a stationary state is not instantaneously achieved or continuously maintained and that the catalyst is either activated or deactivated by secondary reactions. The heat experimentally recorded then differs from the expected value and its experimental determination yields informations on the catalyst activation or deactivation (11).

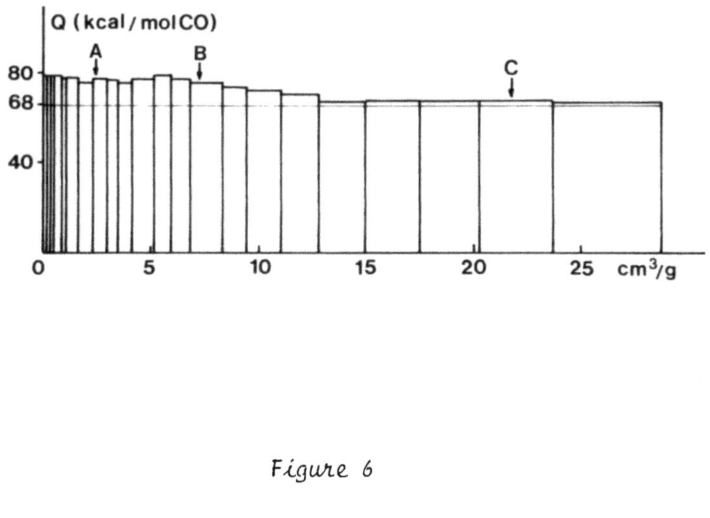

Figure 6

Experimental heats for the combustion of carbon monoxide at 303 K on a sample of gallium-doped nickel oxide as a function of the cumulated volume of reacted gas mixture ($CO + 1/2\ O_2$) (12).

When small doses of the mixture CO + 1/2 O_2 are successively reacted, at 303 K, at the surface of a freshly-prepared gallium-doped nickel oxide catalyst, the heats, initially recorded, (~ 80 kcal/mole) exceed the thermodynamic value for the combustion of CO (~ 68 kcal/mole) (figure 6) (12). The experimental heats however decrease as the number of doses increases until the expected value is obtained. These results indicate that a secondary reaction takes place transiently at the surface of the fresh catalyst and produces the decrease of its activity as shown on the kinetic plots presented for several doses (A, B and C) on figure 7. These plots of the percentage of heat released as a function of time were deduced from the recorded calorimetric curves by application of TIAN equation. The deactivation of the catalyst, thus demonstrated, has been related by direct calorimetric experiments to the irreversible adsorption of part of the reaction product, carbon dioxide, which inhibits the most reactive surface sites (12).

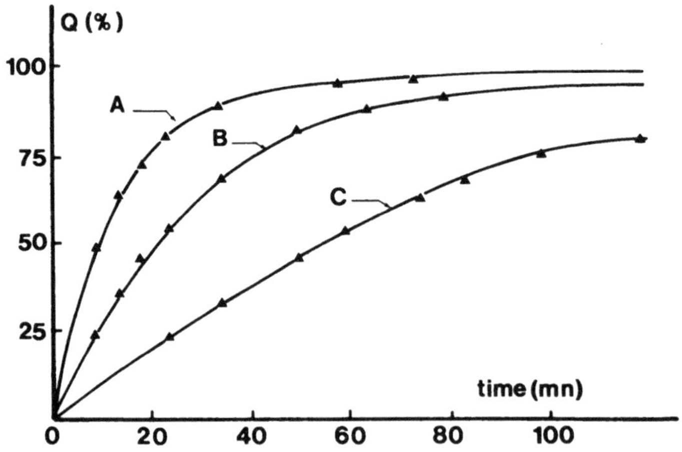

Figure 7

Percentage of evolved heat, as a function of time for the reaction of doses A, B and C in fig. 6 (12).

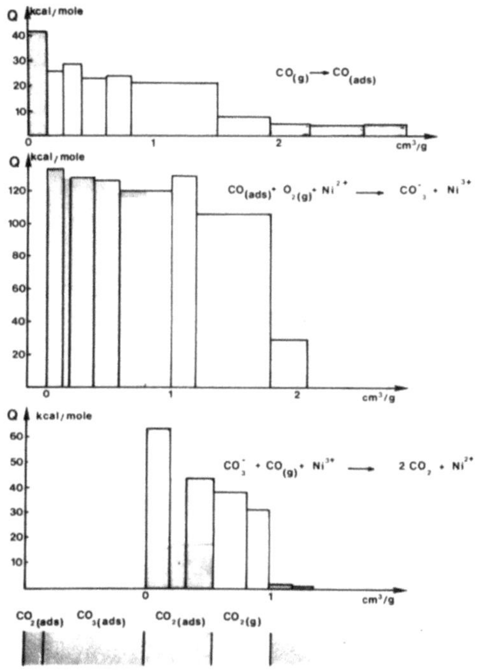

Figure 8

Differential heats of adsorption or interaction measured during the adsorption sequence : CO - O_2 - CO at 303 K on the surface of a nearly stoechiometric nickel oxide. Proposed mechanisms of interaction are indicated on the figure. Shaded areas correspond to the formation of surface species which do not eventually yield the reaction product, $CO_2(g)$, as demonstrated by thermochemical cycles (13).

The same conclusion can also be deduced from more detailed calorimetric studies of the reaction mechanism. The principle of these studies is to adsorb the different reactants in successive sequences and to determine, in each case, the differential heats of adsorption or interaction with preadsorbed species as a function of coverage. Figure 8, for instance, reports the calorimetric results of such a study during which carbon monoxide, oxygen and again carbon monoxide were successively adsorbed at the surface of the same sample of nickel oxide at 303 K (13). A 3 - step mechanism was supposed for the reaction and was compared with the experimental data for different sets of surface sites, by means of thermochemical cycles. Analysis of the data indicate that the most energetic surface sites are liable to be inhibited either by adsorbed carbon dioxide or by adsorbed carbonates, in agreement with the calorimetric results for the reaction itself, presented in figure 6; moreover, some adsorption sites are not active enough to participate in the reaction and finally, the number of sites which may remain active during the reaction is indeed limited. The general picture is therefore in agreement with generally accepted views on the surface distribution of catalytic activity. The merit of this calorimetric method is however to yield some quantitative data about what is generally readily assumed but seldom demonstrated.

Because of the large sensitivity of TIAN-CALVET calorimeters, it may be thought that their operation is tedious and time-consuming. This is true in some cases but not necessarily so. New methods have been recently developed to determine thermokinetics with heat-flow calorimeters which are as rapid and more convenient than more conventional methods in heterogeneous Catalysis (14).

The small glass reactor, presented on figure 9 may be placed in the calo-

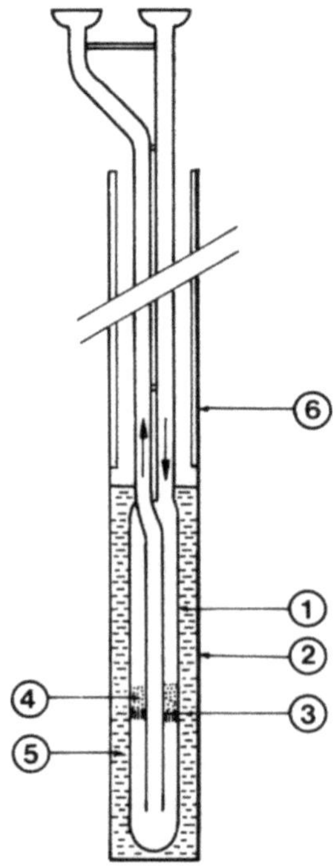

Figure 9

Dynamic reactor and calorimetric cell (14). 1, reactor; 2, calorimetric cell; 3, sintered glass disk; 4, catalyst; 5, silicon oil; 6, insulating tube.

rimeter cell. Thermal contact between the reactor and the sensing elements of the calorimeter is insured by silicon oil. A small amount of catalyst (50 mg) is placed on the sintered glass disk inserted between the tubes which permit the circulation of the reactants in the reactor. The reactor is connected to a classical gas-supply system located outside of the reactor and analysis of reactants and products is achieved gas-chromatographically. Calibration of the calorimeter response for different thermal outputs and different flow rates can be made by placing a small electrical heater in the catalyst bed or by using, as a standard, a catalytic reaction at stationary state. Data reported on figure 10 shows that the proportionality of the calorimeter response to the energy produced in the reactor, measured here in number of moles of product formed per second, is indeed excellent.

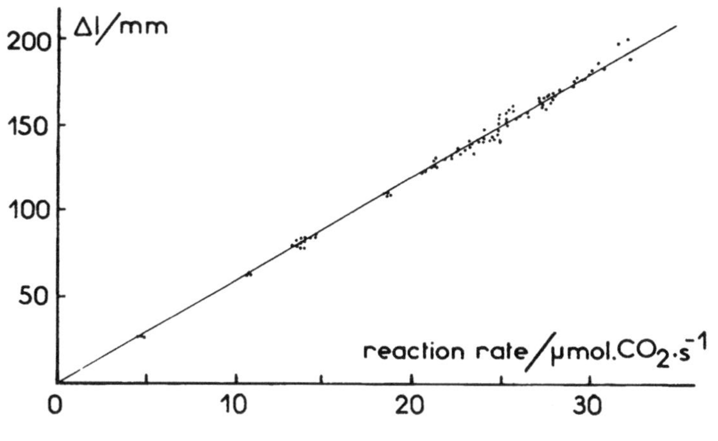

Figure 10

Plot of the measured amount of heat as a function of the quantity of product, detected chromatographically (14).

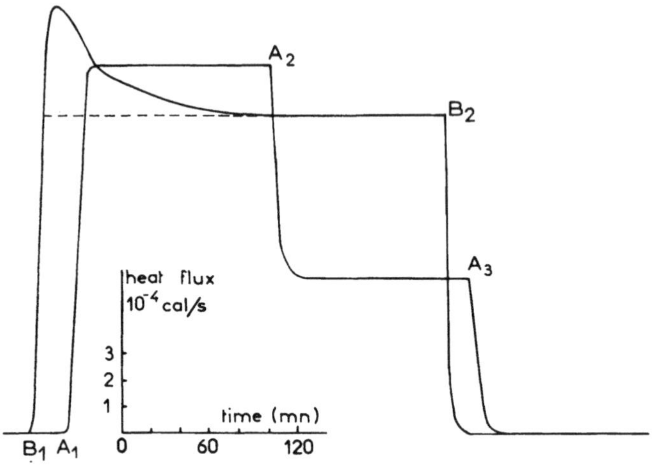

Figure 11

Calorimetric curves recorded during the catalytic combustion of carbon monoxide in the dynamic reactor (14).

Figure 11 shows that, in the case of a catalyst presenting a stable activity (curve A), a stationary state for the reaction is detected by the calorimeter approximately 15 minutes after the introduction of the reac - tants in the carrier gas. Thermogram A presented on figure 11 refers to the combustion of carbon monoxide over a properly aged nickel oxide catalyst at 473 K. First, oxygen diluted in helium was circulated in the reactor. The stable base line then recorded indicates that the catalyst is in equilibrium with flowing oxygen. At time A_1, carbon monoxide was injected in the carrier gas. The plateau, parallel to the base line, which is then recorded indicates that a stationary regime is attained and the ordinate for the plateau is an accurate measure of the catalyst activity. At time A_2, the partial pressure of carbon monoxide in the gas flow was reduced by a factor of 2. The decrease of heat produced, which indicates a decrease of the reaction rate, stops after ~ 15 minutes and a new stationary regime is obtained indicating that the reaction rate varies with the pressure of carbon monoxide in the reaction mixture. In A_3, the flow of carbon monoxide in the carrier gas was discontinued and after some 15 minutes, the thermogram returned to the base line, indicating that no secondary reaction such as the desorption of the reaction product is taking place.

The changes of heat flux recorded during this experiment as the partial pressure of carbon monoxide was varied indicate that this calorimetric method may be utilized to determine rate laws and reaction orders. In the case of the experiment reported on this slide, the experimental conditions were maintained constant for about 2 hrs in order to check the stability of the catalyst activity. When this is demonstrated, an accurate measurement of the activity may be achieved in a much shorter time. The calorimetric determination of reaction orders can usually be done in one day's work.

TABLE 1

Thermokinetic data for the combustion of carbon monoxide at 473 K

flow rate (ml/mn)			heat flux (cal/s)
total	Carbon monoxide	Oxygen	
80	2.6	2.4	1.7
80	2.5	5.5	1.6
80	2.6	7.0	1.7
80	2.7	10.5	1.6
36	0.25	3.2	0.37
37	0.76	3.3	0.89
36	0.90	3.3	0.92
38	1.60	3.5	2.06

The results obtained in the case of CO-combustion over nickel oxide at 200°C are reported on Table 1. It is clear that when the partial pressure of oxygen is varied there is no change in the heat flux produced by the reaction and that the partial order with respect to oxygen is zero. The heat flux, however, changes with partial pressure in the case of carbon monoxide and a first order dependence is demonstrated in agreement with previous results.

The same method is also very effective to detect all secondary processes which may take place, even transiently, at the catalyst surface. Figure 11 also reports thermogram B which was recorded when a mixture of carbon monoxide and oxygen was introduced in the flow of carrier gas composed of pure helium. A thermal peak was first recorded and then a plateau indicating a stationary regime for the reaction. When the introduction of the reaction mixture was discontinued, the calorimetric curve returned to the initial base line. The phenomenon, which explains the peak initially recorded, can be attributed to the irreversible adsorption of oxygen, in excess in the gas mixture, on the surface of the initially nearly stoechiometric oxide. The same results were also obtained previously by introducing successive doses of reaction mixture onto the catalyst (15). But, it took 3 days to perform the experiment in the case of the dosing system whereas 3 hours were sufficient when the flow method was used.

It appears therefore that by placing a dynamic reactor in a heat-flow calorimeter, it is possible to obtain rapidly reliable results on catalyst activities and on rate laws. Moreover, data recording being continuous, it is possible to detect and to study all secondary processes which may take place during the reaction : adsorption, as shown on figure 11, but also catalyst activation or deactivation. Finally, the method may be extended to the study of many reactions since a gas chromatograph is used to analyze the reactants and the products.

The preceding examples have demonstrated that, when a dynamic reactor is placed in the calorimeter, catalyst activity can be easily deduced from the ordinate of the plateau which is recorded when the reaction attains stationary state. Kinetics of transient thermal phenomena, of non-stationary reactions for instance, may also be derived from the calorimetric curves. TIAN equation however shows that the recorded calorimetric curve yields kinetic informations which are distorted by the thermal lag in the calorimeter. Moreover, the use of TIAN equation must be restricted to semi-quantitative kinetic measurements. The assumptions underlying TIAN equation are not completely verified, especially in the case of fast phenomena (16). For TIAN equation to be valid, it is necessary that the surrounding medium remains, at all times, at a constant and uniform temperature. This is not completely true : heat evacuated from the cell transiently accumulates at the internal surface of the surrounding block and changes its temperature. Moreover, temperature gradients must also exist in the calorimeter proper. For these reasons, TIAN equation is only an approximation and, for a quantitative analysis of the data, more complex data-correction must be used to remove the distortion due to thermal lags in the calorimeter.

We have tested and used 3 deconvolution procedures based on Fourier transform analysis, time-domain matrices and state-function theory (17). In all cases, it is assumed that the calorimeter behaves as a linear system with localized constants and we have shown that this is indeed the case with heat-flow calorimeters of the TIAN-CALVET type (18). The principle of the method is then to solve the general equation of linear systems($G(p) = Hp. Fp$)for any particular experiment, once Hp the transfer function has been determined in separate calibration experiments. Digital recording of the data and use of a computer are, of course, necessary. Without entering into any mathematical details, presented elsewhere (17), it may be pointed out that application of state-function theory presents, in our opinion, a number of advantages since, for instance, it allows the on-line correction of the data with a microprocessor. Dirac pulses or Heaviside steps which are needed to calibrate the calorimeter may be produced by Joule heatings but pulse-like adsorption phenomena have also been used for this purpose. Since there is necessarily some uncertainty on the exact location of the heat sources in any particular experiment, it is not possible to completely correct the data. However, we have shown that when the calibration heater or the catalyst occupy a small defined volume of the calorimeter cell, deviations between calibration and actual experiments or for different experiments do not exceed ~ 5 % (19).

The sensitivity of the data correction method was also tested and the li-

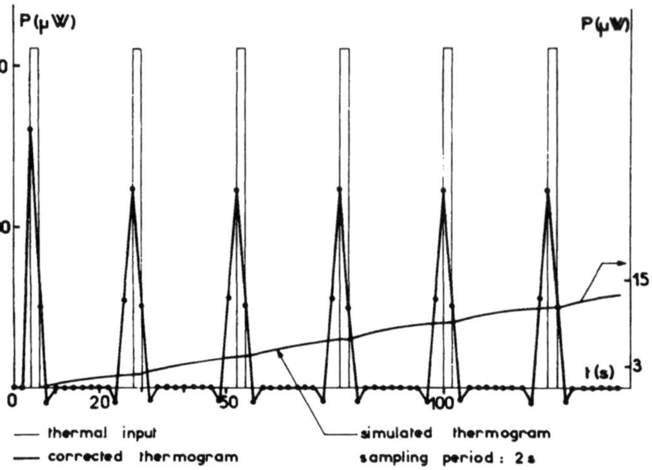

Figure 12

Deconvolution of simulated calorimetric curves (20)

mit of its resolution was determined during simulated experiments (20). Results reported figure 12 indicate that short heat pulses, lasting 1 sampling period (2s) separated by 11 periods (24s) are, after correction, clearly separated, though they are completely blurred on the calorimeter response. Some broadening is however apparent on the corrected thermogram. In the case of less abrupt changes of thermal power in the calorimeter cell, the correction gives even better results. The 60 sec-steps on figure 13 are, for instance, satisfactorily reconstructed by the correction, though the calorimeter response does not give any valuable kinetic information on the thermal process simulated in the calorimeter.

This study has demonstrated that rapid changes in the thermal power produced in the calorimeter can be accurately determined by means of the data correction methods and that reliable kinetic data can be obtained from the thermograms, even in the case of rapid chemical processes. Because of the noise in the recording line, the sampling period which is usually selected

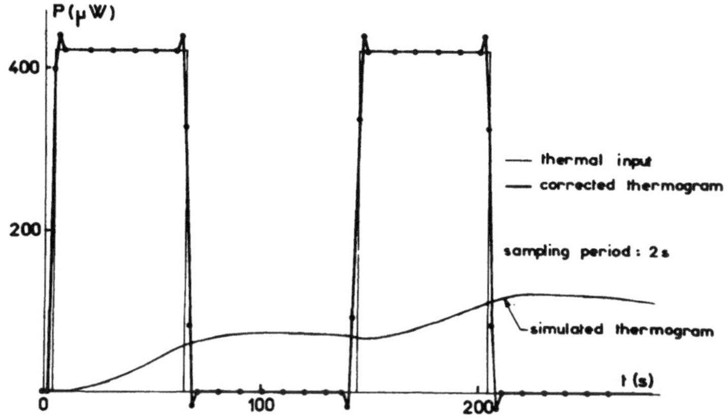

Figure 13

Deconvolution of simulated calorimetric curves (20)

for the digital record of the data is 8 s and not 2 s as in the simulated experiments (20).

Calorimetric data correction methods can be used to obtain kinetic informations on any process producing or absorbing heat : they therefore have a very broad applicability. In the fields of adsorption and heterogeneous Catalysis, they can be employed to study the reactivity or activity spectrum of a catalyst surface.

This particular application is justified by the large sensitivity of the heat-flow calorimetric technique. It is possible to obtain reliable kinetic informations on the adsorption or reaction of a limited number of molecules, covering or reacting on a limited fraction of the surface sites. In the case of an adsorption phenomenon it is thus possible to study the rate law for the adsorption of successive doses. By a progressive poisonning of the active sites, it is possible to determine the activity of different surface fractions.

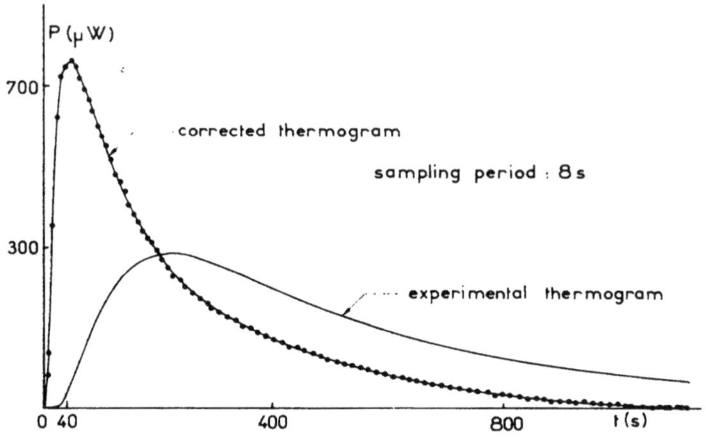

Figure 14

Experimental and corrected calorimetric curves, corresponding to the decomposition at 473 K of a dose of nitrous oxide at the surface of a nearly stoechiometric nickel oxide (21)

In order to test the method, we studied the decomposition of nitrous oxide at 473 K at the surface of a nearly stoechiometric nickel oxide 21). The reaction of a limited volume of N_2O produces, in this case, gaseous nitrogen and adsorbed oxygen, which poisons the sites, where it is adsorbed, for a further reaction. An example of recorded and corrected thermograms is presented on figure 14. The reaction, producing adsorbed oxygen ions, is generally assumed to present a first order dependence with respect to N_2O. If this is the case, the ordinate of the corrected thermogram, at any time, which is proportional to the reaction rate at this time, should be proportional to the area under the thermogram from time t to infinity since this area is proportional to the quantity of reactant remaining in the system at time t. Plots of the heat flux as a function of the thermogram area from time t to infinity are presented on the figure 15. They are indeed linear and they confirm that the decomposition of nitrous oxide to yield adsorbed oxygen is indeed a first order process.

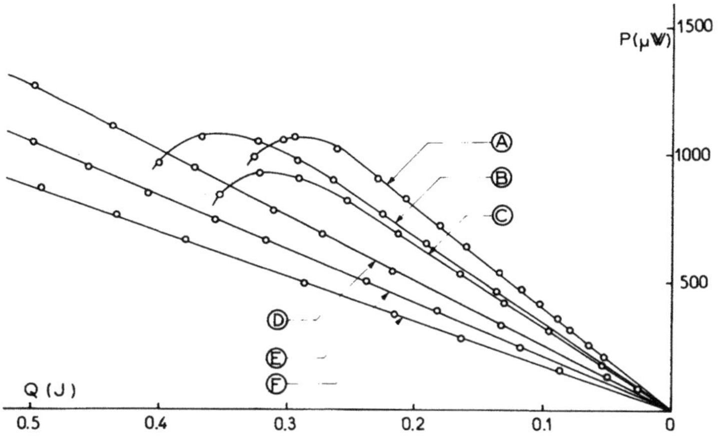

Figure 15

1 st order transforms of the calorimetric data recorded during the reaction of successive doses A to F of nitrous oxide on nickel oxide at 473 K (21)

The first points on any curve are not however on the straight line. They correspond to the very first instants of the reaction (from time zero to 20 - 40 seconds) when the reactant pressure is not stabilized yet near the catalyst bed. As doses are introduced successively onto the catalyst, the slope of the plots which measures the first order activity constant decreases. The different fractions of the catalyst surface therefore exhibit a different activity for this reaction. Values of the activity constant are plotted on figure 16 as a function of the total volume of reactant consumed (filled dots). A two-fold decrease of activity is thus demonstrated from the most active sites to the least active sites.

From the same calorimetric results, it is also possible to calculate the heat produced by the adsorption of oxygen produced during the reaction. This is simply achieved by comparing the experimental heats of reaction to the known enthalpy change for the some reaction in the gas-phase. Heats of adsorption of oxygen are presented as steps on figure 16. It is clear that,

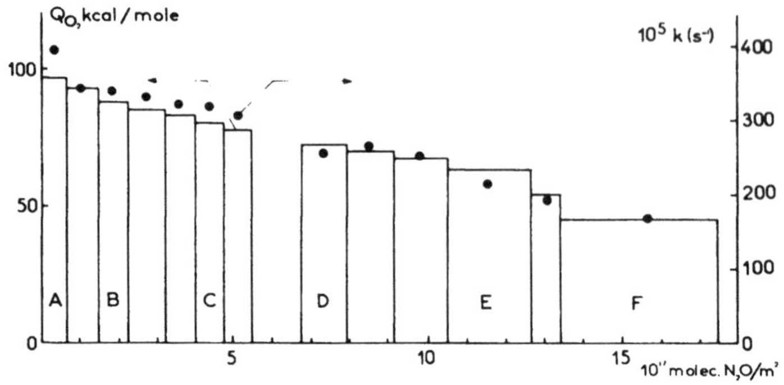

Figure 16

Catalytic activity spectrum of a nickel-oxide catalyst for the decomposition of nitrous oxide at 473 K (filled points) and reactivity spectrum of the same catalyst with respect to oxygen at 473 K (Steps) (21)

in this system, which may be considered as a model system for reduction or oxidation reactions over metal oxide catalysts, the activity of the catalyst surface sites is related to their affinity towards oxygen, measured by the heat of adsorption of oxygen.

Similar experiments can be performed at different temperatures and they allow to determine the distribution of activation energy over the catalyst surface. They can be repeated for different samples and they allow a very precise study of activation and deactivation.

Microcalorimetry can thus provide many illuminating informations about catalytic reactions. In this review, it has been attempted to show that this technique can be utilized to determine i) the reactivity spectrum of catalysts during adsorption, ii) the activity spectrum for the reaction, iii) probable reaction mechanisms, iv) rate laws and v) the secondary processes which cause catalyst activation or deactivation. Calorimetry though still ignored in some laboratories has thus become an essential tool in Adsorption and Catalysis studies, the results of which complement those given by other techniques. In the development of calorimetric techniques, catalytic chemists have not ignored the improvements achieved by calorimetrists working in other fields. Collaboration, contacts, exchange of opinions between specialists of different disciplines are indeed very beneficial and may lead to rapid technological developments. It is hoped therefore that the applications of calorimetry, which have been developed to study catalytic reactions, will suggest possible new uses of this technique in some branches of Life Sciences.

REFERENCES

(1) Calvet, E., Prat, H. : "Microcalorimétrie, Applications Physicochimiques et Biologiques", Masson, Paris, (1956); Calvet, E., Prat, H., Skinner, H.A. : "Recent Progress in Microcalorimetry", Pergamon, Oxford, (1963).

(2) Morrison, J.A., Los, J.M. : The measurement of the thermal properties of gases and vapors adsorbed on solid surfaces. Disc. Faraday Soc., 8, 321, (1950); Chihara, H., Morrison, J.A. : Calorimetric study of some physical phenomena, in "Experimental Thermodynamics", (J.P. Mc Cullough, D.W. Scott, Ed.), Butterworths, London, 1, 537, (1968).

(3) Avgul, N.N., Berezin, G.I., Kiselev, A.V., Lygina, I.A., Muttik, G.G. : The effect of the porosity of graphite adsorbents on the adsorption and heat of adsorption of hexane vapors. Zh. Fiz. Khim., 31, 1111, (1957).

(4) throughout this article, 1 cal = 4.184 J; 760 torr = 101.325 kNm^{-2}.

(5) Dewar, J. : The absorption and thermal evolution of gases occluded in charcoal at low temperatures. Proc. Roy. Soc. London, A 74, 127, (1904).

(6) Kiselev, A.V., Muttik, G.G. : Isothermal calorimeter with constant heat exchange. Zh. Fiz. Khim., 35, 2153, (1961); Dzhigit, O.M.,

Kiselev, A.V., Muttik, G.G. : The measurement of heats of adsorption of pentane and ether by zeolite type 5A by means of an isothermal constant heat exchange calorimeter. J. Phys. Chem., 66, 2127, (1962); Berezin, G.I., Kiselev, A.V., Serdobov, M.V. : An adiabatic differential calorimeter for measuring heats of adsorption on small surfaces. Zh. Fiz. Khim., 36, 2091, (1962); Karnaukov, A.P., Krivousov, M.J., Muttik, G.G., Pankratev, Yu. O. : Calorimeter for measuring the heats of gas chemisorption at 25 to 400°C, calorimeter design. Kin. i Kat., 9, 1369, (1968).

(7) Beeck, O. : Catalysis, a challenge to the physicist. Rev. Mod. Phys., 17, 61, (1945).

(8) Cerny, S., Ponec, V. : Determination of heats of adsorption on clean solid surfaces. Catalysis Rev., 2, 249, (1968).

(9) Cerny, S., Ponec, V., Hladek, L. : Calorimetric heats of adsorption of hydrogen on molybdenum films. J. Catalysis, 5, 27, (1966).

(10) Gravelle, P.C. : Calorimetric measurements in Adsorption and heterogeneous Catalysis. 1st Nat. Conf. on Calorimetry, Pol. Acad. Sci. Pub., (1973).

(11) Gravelle, P.C. : Applications of adsorption calorimetry to the study of heterogeneously catalyzed reactions. Colloq. Intern. Centre Nat. Rech. Sci., Paris, 201, 571, (1972).

(12) Gravelle, P.C., El Shobaky, G., Teichner, S.J. : Determination by means of a Calvet microcalorimeter of the mechanism of carbon monoxide oxidation catalyzed by nickel oxide. VI Gallium-doped nickel oxide. J. Chim. Phys., 66, 1760, (1969).

(13) Gravelle, P.C., Teichner, S.J. : Determination by means of a Calvet microcalorimeter of the mechanism of carbon monoxide oxidation catalyzed by nickel oxide. II Research of possible mechanisms for the oxidation reaction. J. Chim. Phys., 61, 533, (1964); III Research of possible mechanisms for the oxidation reaction. J. Chim. Phys., 61, 625, (1964); IV Study of the oxidation reaction catalyzed by a freshly prepared nickel oxide. J. Chim. Phys., 61, 1089, (1964).

(14) Gruia, M., Jarjoui, M., Gravelle, P.C. : Simultaneous use of a heat-flow microcalorimeter and of a dynamic or pulsed reactor for the study of adsorption or reactions of heterogeneous catalysis. J. Chim.

Phys., 73, 634, (1976).

(15) Gravelle, P.C. : The use of heat-flow microcalorimetry in heterogeneous catalysis research. Proc. 5 th Intern. Congr. on Catalysis, (J. Hightower, Ed.), North Holland, Amsterdam, 65, (1973).

(16) Gravelle, P.C. : Heat-flow microcalorimetry and its application to heterogeneous catalysis. Adv. in Catalysis and Rel. Subj., Academic Press, 22, 191, (1972).

(17) Brie, C., Guivarch, M., Petit, J.L. : Processing of flux calorimeter signal. Proc. 1 st Intern. Conf. on Calorimetry and Thermodynamics, Pol. Sci. Pub., Warsaw, 73, (1971); Brie, C. : Determination of the input applied to a linear system by deconvolution of the output : application to the correction of thermograms in microcalorimetry. Ph. D. Thesis, N° 44, Lyon, (1971); Brie, C., Petit, J.L., Gravelle, P.C. : Application to adsorption and heterogeneous catalysis of a method for the determination of exact thermokinetics. I Thermogram-deconvolution method and experimental arrangement. J. Chim. Phys., 70, 1107, (1973).

(18) Brie, C., Petit, J.L., Gravelle, P.C. : Experimental study of the linearity of a Calvet-type heat-flow microcalorimeter. C.R. Acad. Sci., Paris, 273 B, 1, (1971).

(19) Brie, C., Petit, J.L., Gravelle, P.C. : Application to adsorption and heterogeneous catalysis of a method for the determination of exact thermokinetics. II Verification of the model validity and practical determination of the correction programme. J. Chim. Phys., 70, 1115, (1973).

(20) Point, R., Petit, J.L., Gravelle, P.C. : An improved method for the determination of the input applied to a linear system by deconvolution of the output. Application in thermokinetics. J. Therm. Anal. (in the press).

(21) Gravelle, P.C., Point, R., Petit, J.L. : A microcalorimetric determination of the distribution of active sites at a catalyst surface. J. Catalysis (in the press).

1.2 Ballistic Measurements with the Batch Microcalorimeter

L. Tumerman, R. Zidovezki

Like all other calorimeters based on the heat conduction (or heat leakage) principle the batch microcalorimeter is by nature an instrument with a slow response. The shape of its response to a very short, practically momentary, pulse of heat is shown on Fig. 1. The measured response (voltage) rises approximately linearly up to a time t_{max}, when it reaches its maximal, so-called "ballistic", value V_b and then decays exponentially with the time constant τ. For the instrument we have worked with (LKB Batch Microcalorimeter, Type 10700-2) t_{max}=60sec. and τ=130 sec. Close values of τ have been found for other instruments of this type (1,2).

These instruments may be used, therefore, directly for kinetic measurements only in the case of a relatively slow processes, whose duration is essentially greater than τ. If the reaction is fast in regard to τ, special calculation schemes must be applied to reconstruct the true kinetic curve of the reaction from the recorded voltage-time curve (3-5).

The slowness of the calorimeter may, however, become an asset of the instrument, if only the heat effects of fast reactions are of interest and no kinetic information is required. The response of the batch microcalorimeter (the measured voltage) is proportional to the heat power of the reaction, i.e. to the rate of heat evolution or absorption. The amount of heat Q which was released or absorbed in the course of the reaction can be computed from the area A under the recorded voltage-time curve. Obviously, Q=A/S, where S is the sensitivity of the calorimeter detector expressed in units: V/W. Now it is easy to see from Fig. 1 that in the case of a very short pulse of heat

$$(1) \quad A = \int_0^{\infty} V.dt = V_b(\tau+0.5\ t_{max})$$

and

$$(2) \quad Q = \frac{V_b(\tau+0.5\ t_{max})}{S}$$

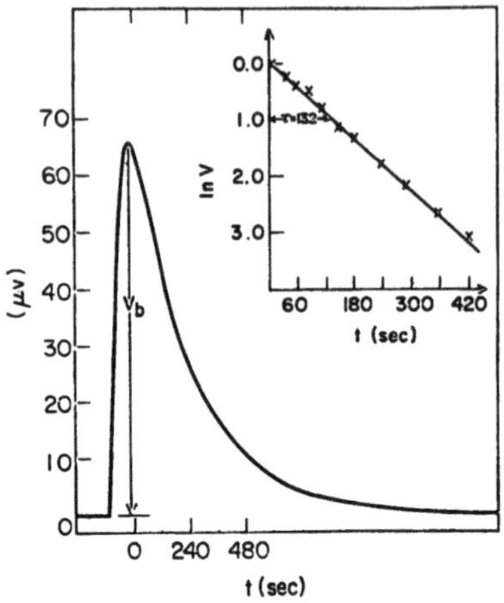

Figure 1. The response of the batch microcalorimeter to a short pulse of heat (I=40ma, t=2sec.). Insert: A semi-logarithmic plot of the decaying branch of the curve.

The heat effect of the reaction is proportional in this case to the ballistic voltage V_b. According to the general theory of measuring instruments we can expect that the same will be true in all cases, when the duration of the reaction is shorter or of the same order of magnitude as τ, since the response of a slow instrument to a short input pulse does not depend upon the shape of the pulse and is always proportional to the "flux" of the measured quantity, i.e. to the area under the recorded response-time curve.

That suggests that for all fast reactions the heat effect Q can be computed simply by multiplying the measured ballistic voltage V_b by the ballistic constant of the instrument

$$(3) \quad B = \frac{Q}{S} = \frac{V_b(\tau+0.5\ t_{max})}{S}$$

For our instrument S=0.061V/W and B=2.66mJ/L.

These conclusions have been checked in a number of electric calibrations experiments in which the values of V_b and A were measured simultaneously as functions of the heat amount Q introduced into the calorimeter by an electric heater. The duration of the heat pulse varied in these experiments between 2 and 40sec. and the values of Q were in the range from 2

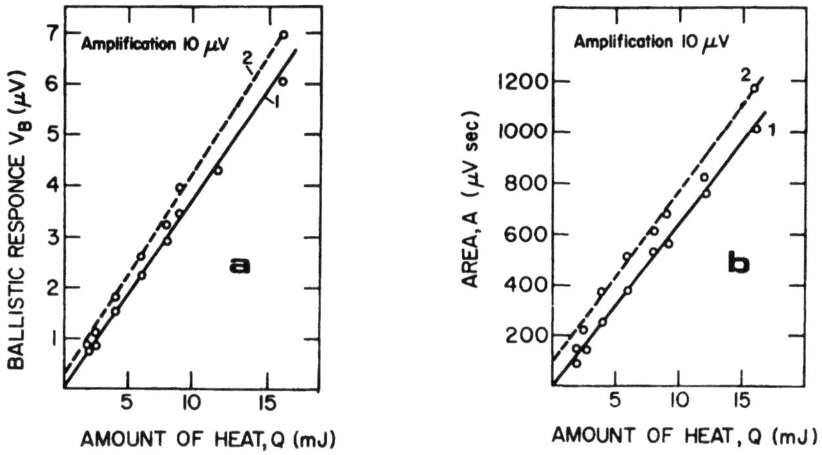

Figure 2a,b. Plot of the ballistic voltage V_b (Fig.2a) and of the area A (Fig. 2b) against Q in electric calibration experiments performed without (line 1) or with (line 2) rotation of the calorimeter unit.

to 200mJ. It can be seen from Fig. 2a,b (line 1) that each of the quantities V and A is proportional to Q. They are therefore proportional to each other. The measured values of the ratio A/V_b and of the ballistic constant of the instrument B are given in Table 1. Small differences between the mean values obtained in various series of calibrations as well as between single measurements in the same series are within the limits of the expected experimental error. Within the same limits the experimentally determined value of B coincides with the one computed above.

In real experiments the reaction is started by rotation of the calorimeter unit. The results must be corrected, therefore, for the "heat of mixing" Q_{mix}, produced by the calorimeter rotation and friction of liquids in the cells. To imitate this situation and determine the value of Q_{mix} we have repeated the above described experiments starting the rotation of the calorimeter simultaneously with or some seconds after the turning on of the calibration current. It has been found that the proportionality between V_b and A exists under such conditions too and that the ratio A/V_b has the same value as in experiments without the calorimeter rotation (Table 1). But on plots of these quantities against Q (Fig. 2a,b) the straight line 2 which are parallel to the line 1 are shifted and do not pass through the

Table 1. Ballistic Parameters of a Batch Calorimeter

Series No.	Number of Experiments	Calibration $B=Q/V_b$ (mJ/µV)	A/V_b (sec.)	A/V_b (sec.)	Calibration & Mixing V_b^o (V)	$K = \frac{Q}{V_b - V_b^o}$ (mJ/µV)
1	12	2.67			0.28±0.12	2.71
2	4	2.62	167			
3	18	2.66	166	186	0.23±0.14	2.56
4	6	2.77	190	206	0.18	2.63
5	10	2.68	179			
6	11	2.60	167	153	0.23±0.18	2.55
7	9	2.62	169	166	0.31±0.19	2.64
	Mean value	2.66±0.02	173±5	178±12	0.25±0.12	2.62±0.03
	σ_{rel}	0.7%	3%	7%	50%	1%

zero point. That means that instead of simple proportionality we have here linear relationships between V_b(or A) and Q:

$$(4) \quad V_b = V_b^o + K.Q \; ; \quad A = A^o + L.Q$$

The most satisfactory set of values of V_b^o and K,as well as A^o and L, has been found by solving the equations (4) by the method of least squares for each set of experimental data. The results of such calculations are presented in two last columns of Table 1, together with the calculated values of the mean error.

One can see that the relative mean error σ does not exceed 1% in determination of the quantities B and K, which do not require the measurements of the area A, whereas the error is essentially greater (between 3% and 7%) in determination of the ratios A/V_b and A/Q. That means that the ballistic voltage V_b can be measured with greater accuracy than the area A. The proposed ballistic method is therefore not only faster and more convenient than the usual method based on measurements of A. It has also the advantage of greater accuracy.

Proportionality between V_b and A was established also in the case of some fast reactions like the reactions of NaCl and HCl solutions dilution and the reactions of the binding of various haptens to corresponding antibodies. For a given reaction the values of A/V_b were always close to those obtained in electric calibrations performed with the same reaction system, so that the heat effects of the reaction $Q^{(r)}$, including the heat of mixing Q_{mix}, may be computed from the relationship:

$$(5) \quad \frac{Q^{(r)}}{Q(cal)} = \frac{A^{(r)}}{A(cal)} = \frac{V_b^{(r)}}{V_b(cal)}$$

For various reaction systems the values of A/V_b may differ from each other and from values given in Table 1, but the differences are small; they did never exceed more than two times the mean square error of measurements.

The main source of errors in measurements of the area A are the poorly controlable shifts of the base line which occur often after the calorimeter rotation. Such shifts influence the area A_{base} which is to be substracted from the measured area A_{run} to obtain the true value of A_{exp}, but they do not interfere with the measurements of the ballistic voltage V_b.

The heat of mixing Q_{mix} can be easily computed from the measured values of V_b^o or A^o, but the relative mean error of such measurements is very high, it exceeds 50%. That shows that the heat effects of repeated mixings are poorly reproducible. This conclusion was supported by direct measurements of the area A_{mix} in some series of repeated mixings with the same reagents in reacted state. The mean value of Q_{mix} obtained in these experiments was close to the value computed from V_b^o, but the dispersion of single results and the mean error were even greater. As a result of all these measurements one can accept for the heat effect of the mixing the value

$$Q_{mix}=(0.8+0.4)\text{mJ}$$

The error of 0.4mJ (or 0.1mcal) can be regarded as the level of the instruments intrinsic "noise" which limits its sensitivity and the accuracy of calorimetric measurements.

References

1. Johansson, A.: Measurement of the time constant of standard LKB microcalorimeters. Science Tools 19, 16 (1972).
2. Stauffer, H.P., Jeanneret, C.: Estimation of characteristic constants of a modified LKB flow calorimeter. Science Tools 19, 17-20 (1972).
3. Calvet, E., Prat, H.: Recent progress in microcalorimetry. Ed. H.A. Skinner, Pergamon Press, London (1963).
4. Hunt, J.B., Ross, P.D., Ginsburg, A.: A calorimetric study of the interaction of Mn^{2+} with glutamine synthetase from Escherichia coli. Biochem. 11, 3716-3722 (1972).
5. Wadsö, I. in: New techniques in biophysics and cell biology, Ed. Pain, R., Smith, B., Vol. 2, John Wiley and Sons Inc., London (1973).

1.3 Development and Properties of Caloric Systems for Substrate Determinations with Immobilized Enzymes

G. Krisam, H.-L. Schmidt

ABSTRACT

For the determination of urea, glucose and uric acid from reaction heats flow systems have been developed by combination of immobilized enzymes and heat detection systems. A device with a column of an enzyme gel in the reaction cell of a heat flow calorimeter permits precise substrate determinations, but is expensive and time consuming. By means of a combination of sensitive thermistors and immobilized enzymes substrates can be determined in the range of 0.05 - 1 µmoles with an accuracy of 0.03 µmoles within 4 minutes; for assay of urine and serum samples only enzymes covalently bound to nylon tubes are suitable. By coupling of enzyme reactions heat yield and therefore sensitivity are enhanced.

INTRODUCTION

The development of high sensitive flow microcalorimeters permitted the analytical application of calorimetry in life sciences. By combination of enzyme catalyzed reactions to calorimetry specific substrate determinations from reaction heat became possible. However since the use of homogeneous systems in flow systems demands high amounts of enzymes and cosubstrates only matrix bound enzymes (1) made possible a step towards routine assay of substrates in calorimetric flow systems.

The reliability of immobilized enzymes in a calorimeter has already been tested by us in an LKB system with a special reaction cell (2). As an example of an irreversible reaction with a known reaction enthalpy the hydrolysis of urea by polyacrylamide-bound urease has been investigated, and urea has been determined in the range of 0.5 - 5 µmoles. The determination of glucose with gel bound glucose oxidase and catalase had been investigated as a model for a coupled enzyme reaction. In this system the reaction heat is dependent on the relation of the activities of the two enzymes, and the best heat yield is obtained with a large excess of catalase activity; in addition saturation of the buffer with oxygen is necessary. Determinations of glucose are possible in the range of 0.03 - 0.5 µmoles with an accuracy of 5 %; time for one analysis is 20 minutes.

Our present investigations aim on the development of inexpensive, rapid and efficient flow systems for routine assays of substrates, which may be realized by means of two principles:
1) Use of thermistors in the detection system to overcome the high time constant of calorimetric heat flow devices, and
2) application of coupled enzyme reactions in order to raise heat yield and to force quantitative substrate conversions.

RESULTS

Thermistor Basing Measuring Device and Electronic Thermostate

The availability of very sensitive thermistors allowed the construction of a simple but sensitive temperature measuring device in which two thermistors are part of a Wheatstone

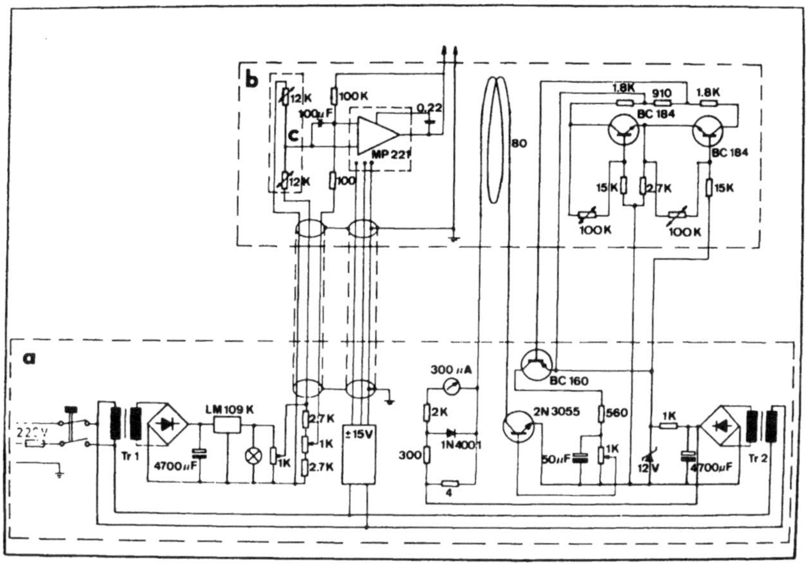

Fig. 1: Thermistor detector and electronic thermostate. a) Control unit, b) thermostate, c) detector.

bridge (fig. 1). Continuous current is supplied to this bridge from an integrated circuit. The difference signal of the bridge is amplified by an Analogic[R] chopper amplifier and transmitted to a compensation recorder with integrator. The detector is located in an electronically regulated thermostate, which contains the temperature sensitive parts of the circuits and the chopper amplifier, too.

The flow through cells for the application of enzyme gels in the thermistor system are made of plexiglass. The buffer solution enters the reaction chamber through a gold capillary bearing the reference thermistor. In the cell for enzyme gels (fig. 2a) the second thermistor is attached to the surface of a gold sheet immersed into the buffer. In the second detector (fig. 2b) the reaction chamber is a nylon tube (10 cm length) with immobilized enzymes (3) connected to two gold capillaries supporting the thermistors. A continuous buffer

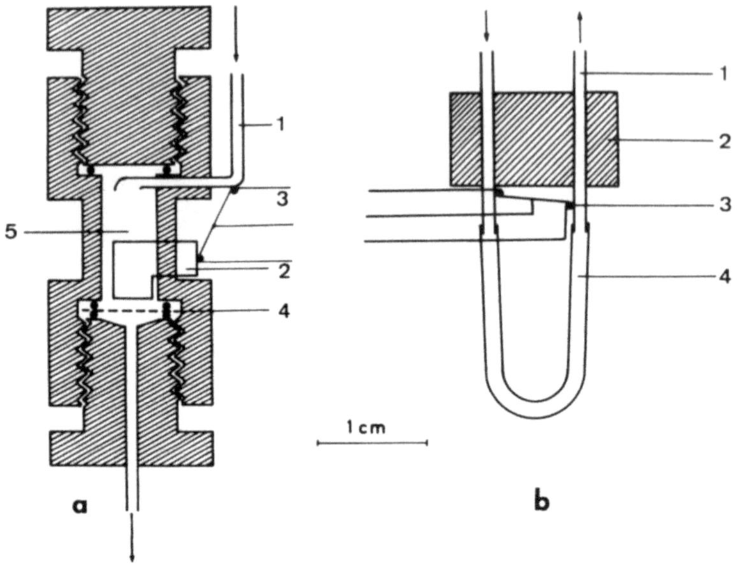

Fig. 2: Reaction cells. a) For enzyme gels: 1) gold capillary, 2) gold sheet, 3) thermistors, 4) sieve, 5) enzyme gel. b) For enzyme bearing nylon tubes: 1) gold capillaries, 2) plexiglass holder, 3) thermistors, 4) nylon tube with immobilized enzyme.

flow (10 ml/h in nylon tubes, 20 ml/h in gel columns) is provided by an Ismatec mp-ge peristaltic pump; substrate dissolved in 50 µl buffer is added by means of an Chromatronix SV 801 valve. Normal reaction temperature is 25 °C.

Substrate Determinations in Biological Samples by Means of Coupled Enzyme Reactions

With the enzyme gel cell glucose in the range of 0.1 - 1 and urea in the range of 0.5 - 5 µmoles can be determined, when the substrates are applied in buffer. However serum or urine caused irreproducible heat effects probably because of adsorption reactions which finally resulted in clogging the enzyme gel cell.

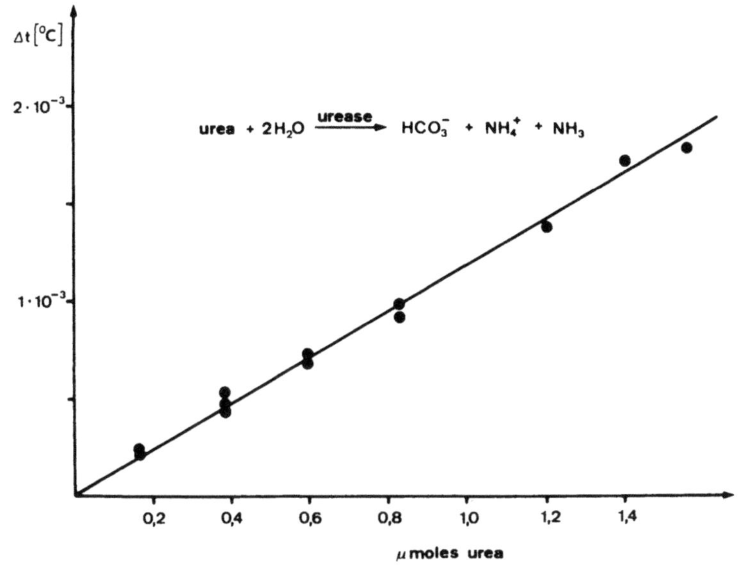

Fig. 3: Thermometric determination of urea through hydrolysis on nylon bound urease. Specific activity: 8 U/m tube; m/10 tris/HCl buffer, pH 7.0.

Therefore detectors with enzymes bound to nylon tubes were used. A tube prepared by direct immobilization of urease lost 80 % of its activity in one day. However when the enzyme was crosslinked with glutardialdehyde in the presence of bovine serum albumine before immobilization (4) an enzyme tube was obtained which only lost less than 10 % of its activity within two months; specific activity of this tube was 8 U/m. Fig. 3 is the calibration curve for urea determinations with a thermistor detector from this urease tube. In spite of the linear relation in the range of 0.1 - 1.5 μmoles urea determinations of the substrate in serum and urine were not yet possible; because of the low molar reaction enthalpy (1.7 kcal/mole), relative errors in the biological medium are too high.

However glucose determinations in biological media were pos-

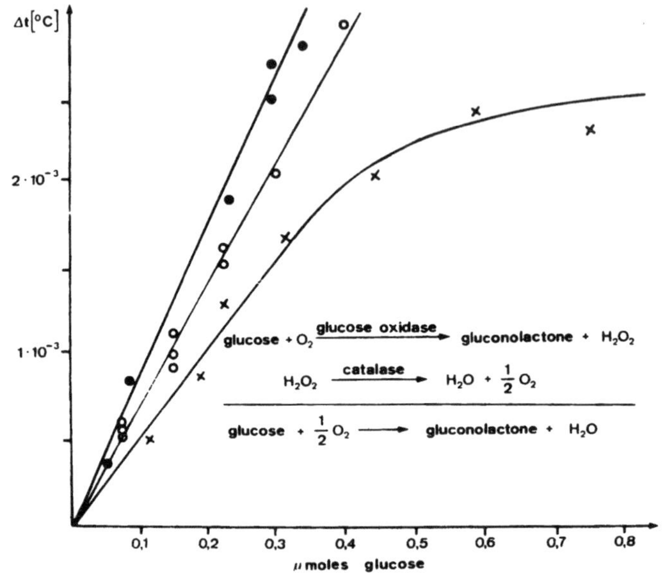

Fig. 4: Thermometric determination of glucose through oxidation on nylon bound glucose oxidase/catalase. Specific activities: +-+-+ 10 U glucose oxidase/m tube, o-o-o 10 U glucose oxidase + 150 U catalase/m tube, •-•-• same, buffer oxygen saturated; m/10 phosphate buffer, pH 8.2.

sible with the coupled enzyme system glucose oxidase/catalase. As already stated with the LKB calorimeter heat yield was depending on the relation of the two enzymes and on oxygen saturation of the buffer (fig. 4). For thermometric assay of glucose in serum and urine 25 µl, diluted with 50 µl of buffer, were applied. While glucose concentrations determined in urine corresponded to values obtained with usual methods (5) within 10 %, values found in serum samples of normal and diabetic subjects exceeded the concentrations of references to a constant amount of 15 - 20 mg %. Yet with a special calibration curve for glucose in serum correct values were obtained (error less than 10 %); 5 minutes were needed for one analysis.

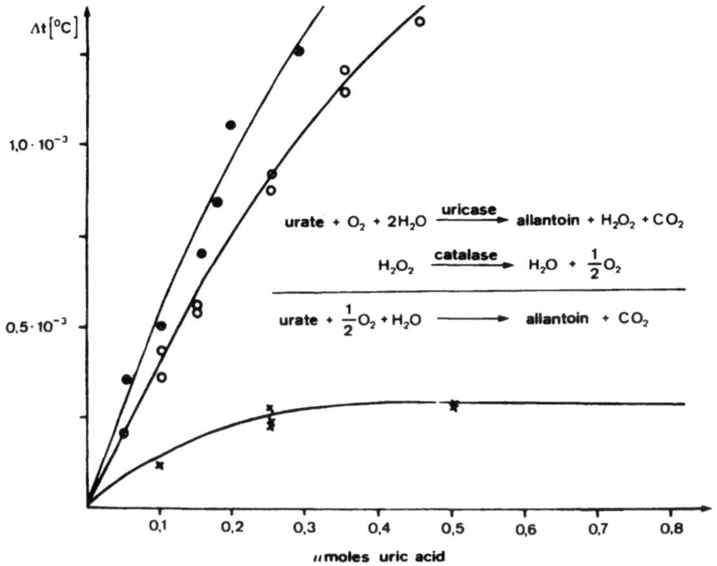

Fig. 5: Thermometric determination of uric acid through oxidation by nylon bound uricase/catalase. Specific activities: +-+-+ 0.1 U uricase/m tube, o-o-o 0,1 U uricase + 250 U catalase/m tube, •-•-• same, buffer oxygen saturated; m/5 borate buffer, pH 9.5.

In general glucose oxidase is already contaminated by catalase, and the effect of an addition of an excess of catalase is not very distinct. Like glucose oxidase uricase is an oxygen dependent H_2O_2 producing flavoprotein; the enzyme is available without catalase contamination. In the thermometric system two nylon tubes were tested, one of which was prepared with uricase, the other with a mixture of uricase and catalase. When urate was applied in the first tube only a small amount of heat was produced (fig. 5), while the coupled system indicated much higher temperature differences for the same amount of substrate. Again saturation of the buffer with oxygen enhanced heat yield. Determinations of uric acid were only possible between 0.05 and 0.5 μmoles; even here the conversion of the substrate was not quantitative, because due to the low specific activity of the starting enzyme material, a specific activity of only 0.1 U/m tube was obtained. After availability of uricase of higher specific activity the same accuracy as in glucose determination will be possible.

As a consequent further development of the coupling of enzyme reactions in order to enhance heat effects systems were investigated which contained three enzymes simultanously bound to nylon tubes. While the enthalpy of hydrolysis of disaccharides ($\sim$1.4 kcal/mole) would not be sufficient for their thermometric determination, sensitivity could be considerably raised, when the hydrolysis was coupled to glucose oxidation:

$$\text{disaccharide} \xrightarrow{\text{hydrolase}} \text{monosaccharide} + \text{glucose}$$

$$\text{glucose} + O_2 \xrightarrow{\text{glucose oxidase}} \text{gluconolactone} + H_2O_2$$

$$H_2O_2 \xrightarrow{\text{catalase}} H_2O + 1/2\ O_2$$

After preliminary experiences with saccharase not only determinations of different disaccharides seem to be possible but also the continuous control of their concentration.

CONCLUSIONS

In general flow systems for substrate determinations basing on calorimeters are more precise but less rapid and by far more expensive than those using thermistor detectors. However these may be useful in semiquantitative routine assay or in concentration control of substrates. A disadvantage of the thermistor system is the dependence of its response on the buffer speed; therefore the use of pumps with highly constant output is essential.

Especially by use of nylon bound enzymes determinations of substrates in biological liquids become possible. The lower limit of detection in the thermometric system being about 100 μcal, 0.05 μmoles of a substrate can be detected, when the reaction enthalpy is in the range of 5 kcal/mole. By coupling of reactions a 10-fold enhancement of sensitivity can be obtained. Further development of calorimetric detection systems is mainly depending on the availability of enzymes with high specific activity and on sophisticated methods for their immobilization.

This paper is part of the thesis of G. Krisam, Freising 1976

REFERENCES

(1) Johansson, Å., Lundberg, I., Mattiasson, B., Mosbach, K.: The application of immobilized enzymes in flow calorimetry. Biochim. Biophys. Acta 304, 217 - 221 (1973).

(2) Schmidt, H.-L., Krisam, G., Grenner, G.: Microcalorimetric methods for substrate determination in flow systems with immobilized enzymes. Biochim. Biophys. Acta 429, 283 - 290 (1976).

(3) Morris, D. L., Campbell, J., Hornby, U. E.: A chemistry for the immobilization of enzymes on nylon. Biochem. J. 147, 593 - 603 (1975).

(4) Avrameas, S., Ternynck, Th.: The cross-linking of proteins with glutaraldehyde and its use for the preparation of immunoadsorbents. Immunochemistry 6, 53 - 66 (1969).

(5) Bergmeier, H. U.: Methoden der enzymatischen Analyse, 2nd ed., Verlag Chemie, Weinheim/Bergstr. 1970.

1.4 The Application of a Heat Transport Model for the Assessment of Regional Blood Flow and Regional Metabolic Heat Production in Dog Myocardium

W. Müller-Schauenburg, H. Benzing

The classical access to tissue energy supply is the measurement of two global parameters of complete organs:
i) The arterial-venous difference of oxygen content and
ii) Complete organ blood flow.
An attempt to measure regional heat production was given by GRAYSON and coworkers (1, 2). They answered the permanently challenging question of the interpretation of stationary tissue temperature difference from aortic input temperature, by comparing the temperature increase caused by locally heating the tissue with known energy input into a tiny probe. This approach resulted only in arbitrary heat production units. On the basis of measurement of regional blood flow, we extended our approach (3) to obtain heat production in more quantitative terms in well defined dimensions of mW/ml tissue (4, 5).

Our approach is not that of closed-system calorimetry, it is not even a measurement on a complete organ but rather a determination of regional heat production in some 0.1 ml volume of tissue. Besides the heat production, the regional blood flow is considered as a second parameter describing regional energetic state of a normal or damaged tissue area, e.g. during myocardial infarction.

The basis of our evaluations are the recordings (fig. 1) of the following parameters:

*Supported by the Erwin-Riesch-Foundation

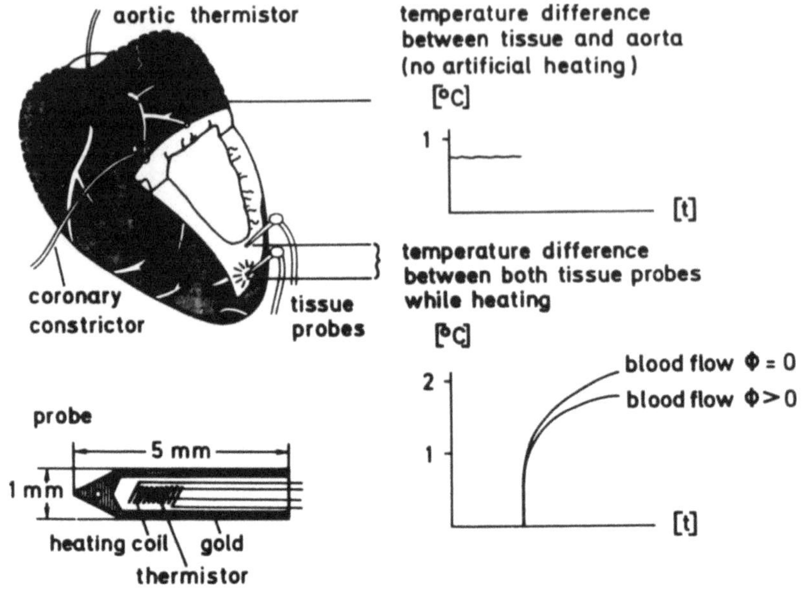

Fig. 1. Left top: The alternately heated probes implanted into the myocardial wall and the balloon occluder for intermittent coronary occlusion. Left bottom: An enlarged scheme of the probe construction.
Right top: Difference between stationary metabolic tissue temperature and aortic temperature. Right bottom: Temperature step response to heating current. Two different curves of zero blood flow and non-zero blood flow are superimposed, the former being used as a reference

i) Local myocardial tissue temperature

ii) Aortic blood temperature as input temperature to the coronary circulation.

Two different types of heat are involved in the measuring process (fig. 2):

First, there is endogenous heat produced by continuously working heart muscle. The arterial blood enters the myocardial tissue at aortic temperature, equilibrates with the tissue temperature by taking up metabolic heat continuously and transporting it to the central venous blood pool. This continuous heat clearance by the blood is supplemented by some heat conduction mainly to the endocardial surface where the tissue

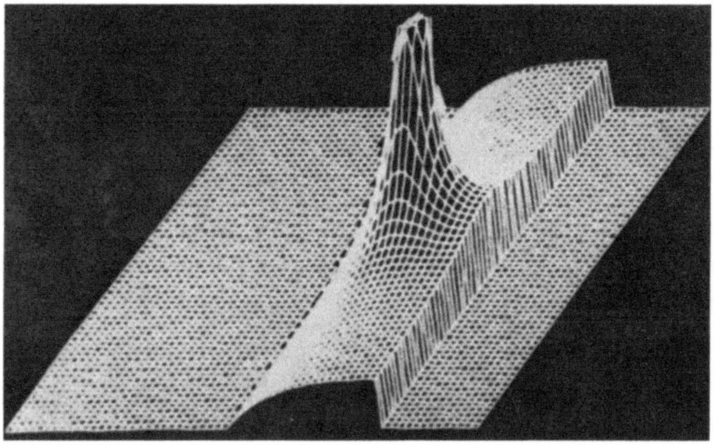

Fig. 2. Schematic cut through a computed temperature field in the left ventricular wall of the myocardium. The endocardial surface is kept at constant temperature, the lung surface, in the first approximation, is assumed to act as heat insulator. The temperature is plotted in the third dimension. The metabolic heat produced all over the myocardium yields temperature increment of slightly more than half a degree centigrade above the ventricular blood temperature

temperature is kept constant at blood temperature level, i.e., the aortic level.

Besides this continuous process of metabolic heat clearance, an artificial heat is produced intermittently in the heating coil of our probe for the determination of blood flow.

All types of heat transport within the tissue are analyzed on the basis of the same model (cf. 6, 3). The model includes

- heat conduction as a diffusion process and
- heat convection, e.g. the transport by blood flow.

The description of heat transport by blood is not concerned with the vessel structure or the microscopic structure of blood flow velocity vectors but it looks at the blood simply as a means of supplying and clearing tissue area. This leads phenomenologically to the dimension

$$\frac{\text{ml blood}}{\text{ml tissue} \cdot \text{min}}$$

without reference to any direction.

For the determination of regional blood flow one can separate heat convection and heat conduction: One does not consider the temperature time course itself but its slope

$$du_\Phi/dt = du_0/dt \cdot \exp(- \Phi \cdot t/\lambda)$$

where the temperature difference u_0 refers to zero blood flow. The reference curve is first estimated from measurement in gelatine and used on-line for computation of blood flow values. If necessary, these are later corrected using post-mortem reference curve. Evaluations are done on a desk calculator HP 9810 A.

The first step in determination of regional heat production confines to that part which is cleared by blood flow. The approach to this convective part is straigthforward; when entering the tissue the blood is warmed up by u ^oC. These are converted into Joules via the specific heat per blood volume ρc_{blood} (Joule $\cdot$ $^oC^{-1}$ $\cdot$ ml^{-1} blood). Thus $\rho c \cdot u$ are the Joules taken up per ml of blood. On multiplication by ml of blood flow per ml of tissue and min one obtains heat in Joules taken·up by 1 ml of tissue in 1 min.

This convective part of metabolic heat production has to be corrected for heat diffusion to the tissue surface (fig. 3). This correction depends heavily on blood flow. For a normal blood flow of 0.7 ml blood $\cdot$ ml^{-1} tissue $\cdot$ min^{-1} the correction amounts to some 15 p.c. With increasing blood flow it becomes less important. For small blood flow the convective part vanishes, and the diffusion process dominates in the direction mainly to the ventricle and less to the lung surface. Geometric assumptions, partially included in fig. 2, from which cor-

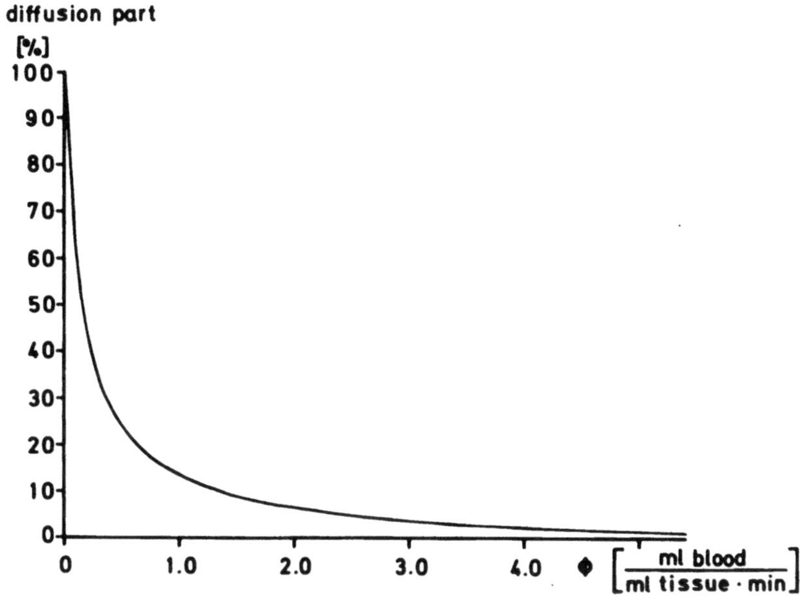

Fig. 3. Correction for diffusion calculated from the model temperature field, described in fig. 2

rection for diffusion is derived are based on approximating the left ventricular wall by a sphere of inner radius 1.5 cm and thickness of 1 cm. The temperature is assumed to be kept constant at the ventricular surface due to good mixing of ventricular blood. At the lung surface thermal isolation is used as a first approximation.

Fig. 4 shows the results of a single experiment with an anesthetized dog. During systemic stress (hypoxemic hypoxia) and at rest the measured values at the three measuring sites differed only accidentally. Starting with intermittent coronary occlusion (60 min) the differences were due to regional tissue damage. Fig. 4 demonstrates the different reactions to reduced oxygen content in the inspired air. The white columns demonstrate normal reactive increase of blood flow during hypoxia. The tissue damaged by the preceding occlusion cannot react in this way. Here mere recording of tissue temperature difference is completely misleading: The temperature difference does not

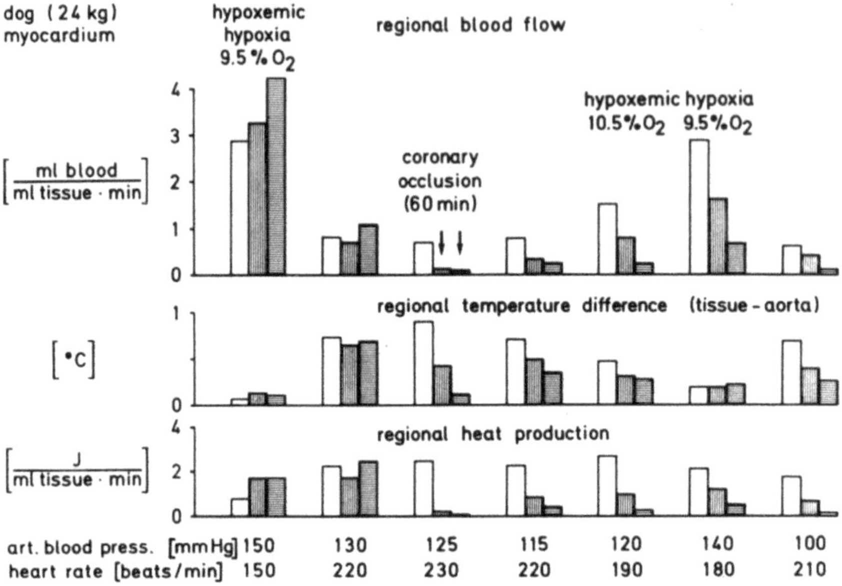

Fig. 4. From top to bottom: Regional blood flow, temperature difference between tissue and aorta, regional heat production and control parameters. There are three tissue measuring points; white (hatched) columns = outside (inside) the region damaged during the experiment by intermittent coronary occlusion (cf. arrows). Time course without scale from left to right: 1. hypoxemic hypoxia, 2. basic state, 3. coronary occlusion, 4. control after release of occlusion, 5. and 6. hypoxemic hypoxia (10.5 and 9.5% O_2), 7. final control

differ in normal and damaged tissue; only when regional blood flow and regional heat production are considered, it is possible to estimate the energetic situation of the tissue.

The table shows mean values of regional heat production and regional blood flow in different experimental states of the tissue. The results are derived from 30 experiments. They have been compared (11 experiments) with classical energy turnover method based upon global O_2 uptake.

$$\left(\text{regional heat production}\right) = \left(\text{regional blood flow}\right) \left(\text{oxygen uptake}\right) \left(\text{energy equivalent}\right)$$

$$\left[\frac{\text{Joule}}{\text{ml tissue}\cdot\text{min}}\right] = \left[\frac{\text{ml blood}}{\text{ml tissue}\cdot\text{min}}\right] \left[\frac{\text{ml } O_2}{\text{ml blood}}\right] \quad 20\left[\frac{\text{Joule}}{\text{ml } O_2}\right]$$

Thus the complete organ blood flow has been replaced by regional blood flow in the classical formula.

Table. Regional blood flow, regional heat production and O_2-uptake control ($\bar{x} \pm s$). n: number of measuring points in the myocardium. 1 Joule/s = 1000 mW (= 0.24 cal/s)

state	regional blood flow	regional heat production (convect.part)	O_2 control
	$\frac{\text{ml blood}}{\text{ml tissue}\cdot\text{min}}$	$\frac{\text{mW}}{\text{ml tissue}}$	$\frac{\text{mW}}{\text{ml tissue}}$
basic state	0.65 ± 0.30 (n = 96)	27 ± 13 (n = 96)	35 ± 12 (n = 33)
hypoxemic hypoxia 10.2±0.5% O_2	1.58 ± 0.77 (n = 52)	35 ± 15 (n = 52)	29 ± 9 (n = 22)
coronary occlusion	0.31 ± 0.28 (n = 28)	6 ± 6 (n = 28)	17 ± 5 (n = 5)

The mean values of regional blood flow quoted above agree with previous results (7) reflecting the fact that the blood flow does not drop to zero values (8) in dog experiments because of collateral circulation (9).

The results on heat production (second and third column of the table) differ significantly. The mean values become more similar if the correction for diffusion is included. This correction is necessary particularly for cases of low blood flow but it is somehow arbitrary depending on the temperature field assumed and on the implantation depth of the probes (cf. fig. 2 and 3).

The method is suitable for organs where heat diffusion is not dominant, either because of the large distance to the organ surface or due to high blood flow. Since the tissue is damaged by introducing probes it is not possible to carry out routine clinical studies. By means of animal experiments it is possible to test methods for the therapeutic improvement of energetic conditions in localized areas of damaged tissue as in coronary disease.

References

1. GRAYSON, J., COULSON, R.L., WINCHESTER, B.: Internal calorimetry-assessment of myocardial blood flow and heat production. J. appl. Physiol. 30, 251-257 (1971).
2. GRAYSON, J., SCOTT, C.A., MORRISON, C.J.: Oxygen utilization and coronary vascular reserve in the ischemic myocardium following acute coronary occlusion in the dog. Microvasc. Res. 11, 181-193 (1976).
3. MÜLLER-SCHAUENBURG, W., APFEL, H., BENZING, H., BETZ, E.: Quantitative measurement of local blood flow with heat clearance. Basic Res. Cardiol. 70, 547-567 (1975).
4. BENZING, H., MÜLLER-SCHAUENBURG, W., BETZ, E.: Determination of regional heat production and regional blood flow in dog myocardium. Dtsch. Physiol. Ges. in cooperation with the Österr. Physiol. Ges. and the Österr. Biophysik. Ges. 45th meeting (autumn meeting) from Sept. 23-26, 1975 in Wien. Pflügers Arch. Suppl. to Vol. 359, R 147 (1975).
5. BENZING, H., MÜLLER-SCHAUENBURG, W., BÜHLER, M.: Bestimmung von regionaler Wärmebildung und Durchblutung im Myokard narkotisierter Hunde. Biomed. Technik 21, Ergänzungsband, 229-230 (1976).
6. PERL, W.: Heat and matter distribution in body tissues and the determination of tissue blood flow by local heat clearance methods. J. theoret. Biol. 2, 201-235 (1962).
7. BENZING, H., WAHL, S.H., BENDER, H.P., RABE, M.: Quantitative local blood flow changes in the insufficiently supplied dog myocardium, measured by means of the heat-clearance method. 7th Europ. Conf. Microcirculation, Aberdeen 1972, part I, Bibl. anat. 11, 139-144 (1973).
8. RIVAS, F., COBB, F.R., BACHE, R.J., GREENFIELD, jr., J.C.: Relationship between blood flow to ischemic regions and extent of myocardial infarction: Serial measurement of blood flow to ischemic regions in dogs. Circulat. Res. 38, 439-447 (1976).
9. SCHAPER, W.: The collateral circulation of the heart. In: Clinical Studies. Vol. 1. North-Holland Publ. Comp. Amsterdam-London. American Elsevier Publ. Co. Inc., New York, 1971.

2. Biochemical Calorimetry

2.1. Hydration and Thermal Stability of α-Lactalbumin. A Calorimetric Study

M. Rüegg, U. Moor, A. Lukesch, B. Blanc

SUMMARY

The effect of hydration upon the thermal denaturation of α-lactalbumin has been studied by differential scanning calorimetry. A strong dependence of the denaturation temperature (T_D), the enthalpy (ΔH_D) and entropy changes (ΔS_D), and the sharpness of the transition peak ($\Delta T_{1/2}$) was observed in the water content range of 0 to about 0.3 g/g (region of primary hydration water). In the water content range of 0.3 to about 0.8 g/g (region of secondary hydration water), smaller changes of the thermodynamic parameters of the thermal denaturation were caused by water uptake. Upon further dilution of the protein in water, ΔH_D remained constant (approx. 59 kcal/mole), but T_D and $\Delta T_{1/2}$ increased from 60.2°C and 7.5°C in 40-20% solutions (wt% protein) to 63.7°C and 10.7°C in 10-3% solutions, respectively. Reversibility of the reaction in these solutions was 60-70%. The thermal transition observed in water does not meet the requirements for a purely two-state process. Different results are obtained in electrolyte solutions. For example, in a solution of phosphate buffer, ΔH_D and T_D increase. This is explained in terms of the conformation stabilizing effect of phosphate ions. Sucessive measurements of the same sample in phosphate buffer showed that the transition was 80-90% reversible for the particular thermal history involved in the study.

INTRODUCTION

Studies on the thermal denaturation of milk proteins are of importance in understanding both the structure and stability of these proteins as well as the changes in the properties of milk and milk products during heat treatments. With the development of highly sensitive thermoanalytical techniques, quantitative measurements of the thermal denaturation of proteins have become feasible with greater experimental ease than with other classical methods. Differential scanning calorymetry (DSC) is one technique which offers the possibility of direct measurement of the most important thermodynamic parameters characterizing heat-induced transitions (1,2). We used this technique for studying the thermal properties of the β-lactoglobulin-water system and for other protein-water systems (3,4). In these systems the thermally induced conformational changes of the proteins were irreversible (3) or only partially reversible (4). It was considered useful to extend this research and to study another important whey protein which undergoes reversible thermal denaturation (5), namely α-lactalbumin.

As in the case of β-lactoglobulin the thermal denaturation of dilute α-lactalbumin solutions has been extensively studied by various techniques such as immunodiffusion (6), spectrophotometry (5), pH-potentiometry (7) and, by measuring the amount of heat-induced centrifugable precipitate (8), but, to our knowledge, there are no reports on the influence of hydration upon the thermal stability of this protein. Also, many published studies on thermal denaturation did not take into account the considerable degree of renaturation of α-lactalbumin after heat treatments.

The aim of the present work was to study the influence of hydration on the heat denaturation of α-lactalbumin which is observed in the temperature range of about 50-70°C, and to compare the results with those obtained from similar measurements of other globular proteins.

EXPERIMENTAL

Materials

α-lactalbumin was prepared from pooled bovine milk of Simmenthal cows by gel filtration followed by ion exchange chromatography according to the method of Armstrong et al. (9). A molecular weight of 14,176 was chosen for the calculation of molar concentrations (10). Purity of the preparations was checked by polyacrylamide gel electrophoresis and by analysis of amino acid composition (10). The purified solutions were dialyzed against distilled water and lyophilized. The water content of the samples used for calorimetric measurements was adjusted either by adsorption of water vapor at appropriate vapor pressures or, for the highest water contents, by directly adding distilled water to the dry preparations.

Calorimetry

Calorimetric measurements were made using a Perkin Elmer model DSC-2 differential scanning calorimeter. The experimental procedure has been described in detail previously (3). Samples of hydrated α-lactalbumin were packed and sealed in standard volatile sample pans (Perkin Elmer) and scanned at a rate of 5°/min until the denaturation process was complete. The heat capacity of the reference cell was balanced using an appropriate amount of water. Except for dilute solutions the water content of the samples was determined after each experiment by drying the punctured sample pans to constant weight over phosphorus pentoxide. Protein concentration in dilute solutions was determined spectrophotometrically at 280 nm in 0.1 M phosphate buffer of pH 6.9 using $E^{1\%}_{1\ cm}$ = 20.6 (11).

Water Sorption Data

Water sorption by α-lactalbumin in the relative water vapor pressure range (p/p_o) of 0.05 to 0.98 was determined at $25^{o}C$ using an isopiestic technique. The apparatus and technique have been described in detail elsewhere (12,13). The equilibrium water contents in the p/p_o range of 0.05 to about 0.90 were obtained with an accuracy of ± 0.5%. At higher p/p_o values the experimental precision was somewhat lower, because of the steepness of the sorption isotherm in this p/p_o region (approx. $\pm$ 1%).

RESULTS AND DISCUSSION

Hydration of α-lactalbumin

Many experiments have suggested that hydrated globular proteins are surrounded by different shells or layers of water (3, 14, 15 and others). The water existing near the protein surface has thereby been characterized by different degrees of interaction with the solute. Using the concept of primary and secondary hydration water, as opposed to bulk water (4), we may divide the water sorption isotherm for α-lactalbumin into different segments:

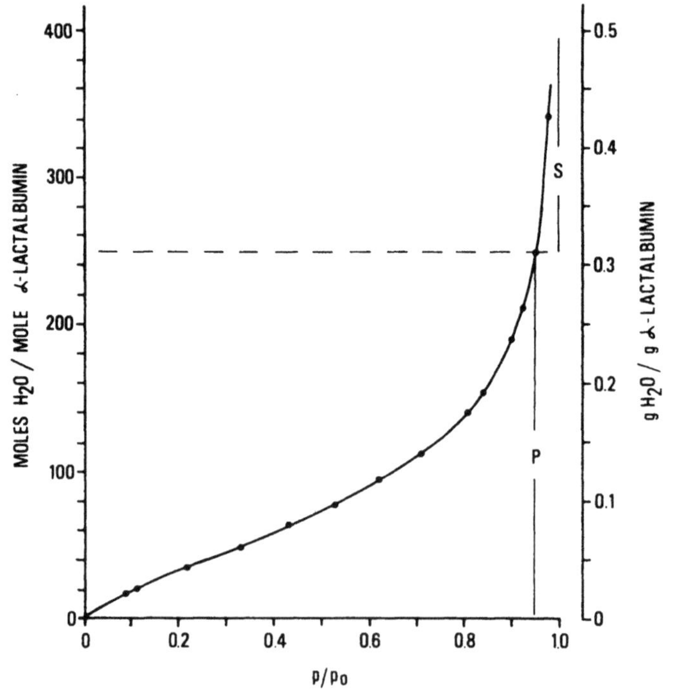

Fig. 1

Water sorption isotherm for α-lactalbumin at 25°C. P: primary hydration water; S: secondary hydration water; p/p_o: relative water vapor pressure.

1) In the water content range from dryness to about 250 mole/mole (0.31 g/g) primary hydration water (P) is absorbed. This strongly bound water cannot participate in a normal ice lattice without rupture of hydrogen bonds with the protein and thus does not freeze at temperatures as low as $-70^{\circ}C$. Various studies have suggested that this strongly modified water has particular physical properties which allow a further subdivision (14, 16, 17). For α-lactalbumin primary hydration water corresponds to the amount of water absorbed in the p/p_o range from 0 to about 0.95. It is interesting to note that the amount of strongly bound primary hydration water (0.31 g/g) may be compared with that observed for a number of globular and fibrillar proteins. Albumins (18), β-lactoglobulin (3), caseins (19), collagen (16) and keratin (20) all revealed an amount of non-freezable water around 0.3 g/g. For α-lactalbumin this amount corresponds to about 2 molecules of water per amino acid residue.

2) Above the critical water content of 250 ± 16 moles H_2O/mole protein, to about 700 mole/mole, thermograms reveal freezable water with temperature and heat of fusion different from ordinary water. A melting thermogram of this secondary hydration water (S) is shown in Fig. 2. As judged from a measurable heat absorption, fusion of this modified water started in the vicinity of $-30^{\circ}C$ and was complete around $-1^{\circ}C$. The complex shape of the melting curve indicates the presence of different types of secondary hydration water.

3) At water contents higher than approximately 700 mole/mole water is absorbed which cannot be distinguished by calorimetric methods from the bulk liquid, i.e. heat and temperature of fusion of the water associated with the protein are, within the experimental error, equal to the corresponding values for pure water.

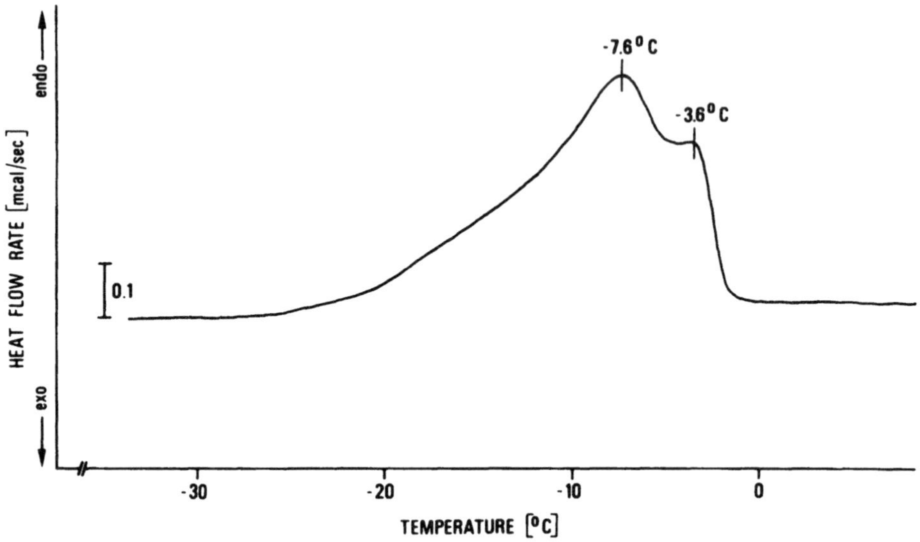

Fig. 2
DSC thermogram of secondary hydration water of α-lactalbumin. Sample weight: 11.767 mg; water content: 409 moles H_2O/mole protein; sample equilibrated at -25°C during 14 h before DSC scan at 2.5°/min and full scale sensitivity of 1 mcal/sec.

Dependence of Thermostability on Hydration

In the β-lactoglobulin-water and collagen-water systems the changes of the state of water absorbed upon stepwise hydration were apparently accompanyied by changes in the heat stability of the proteins (3,4). In the case of α-lactalbumin the same phenomenon is observed. Fig. 3 shows typical denaturation thermograms for α-lactalbumin obtained at different levels of hydration of the protein. The following thermodynamic data were measured in such DSC thermograms and plotted as a function of water content in Fig. 4: the enthalpy of denaturation ΔH_D as obtained from the peak area, the temperature of denaturation T_D at maximum heat absorption, and the half width of the endothermal transition $\Delta T_{1/2}$. Also included in Fig. 4 are the entropy changes ΔS_D, which were calculated from T_D and ΔH_D for

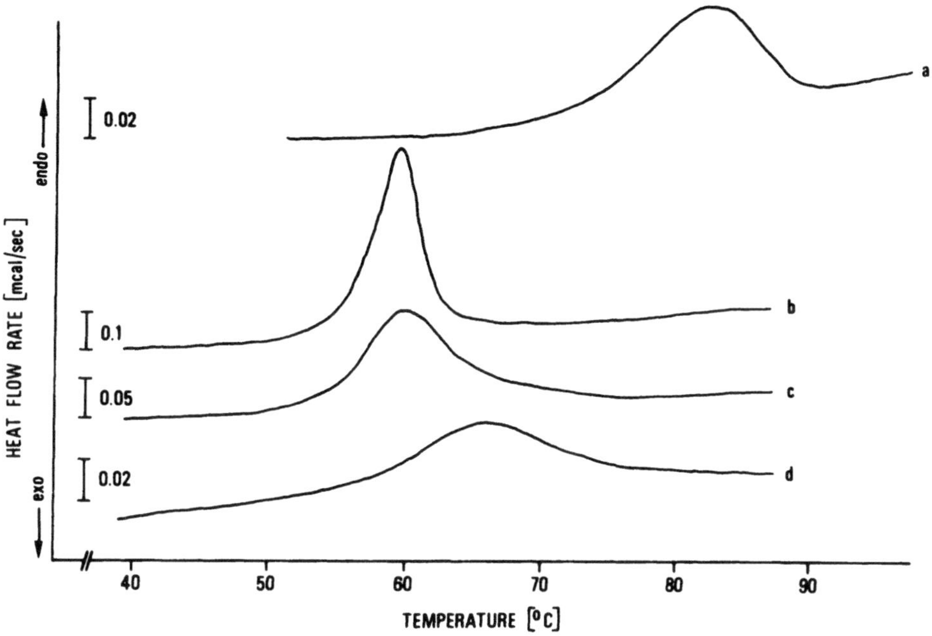

Fig. 3
DSC thermograms of α-lactalbumin in the temperature range 30-100°C. a)-c): α-lactalbumin-water system; water contents (mole/mole) and sample weights (mg) are 44/3.298, 393/11.813 and 1857/10.147, respectively. d): 6.0% α-lactalbumin in 0.1M phosphate buffer solution of pH 6.9. Heating rate is 5°/min.

$\Delta G = 0$ at the transition point ($\Delta S_D = \Delta H_D / T_D$). The predenaturational changes which have been observed by Privalov et al. (21) to occur before the effective thermal transition have not been considered in this study. These predenaturational processes are not accompanied by sharp changes of physical parameters. A linear increase of heat capacity takes place with the rise in temperature only.

All of the four functions plotted in Fig. 4 show a marked dependence on the degree of hydration of the protein. In the water content range from dryness to about 700 mole/mole the experimental points for the denaturation temperature, T_D, apparently follow an exponential curve. This indicates that the T_D values follow an equation similar to that found by Flory and Garrett for

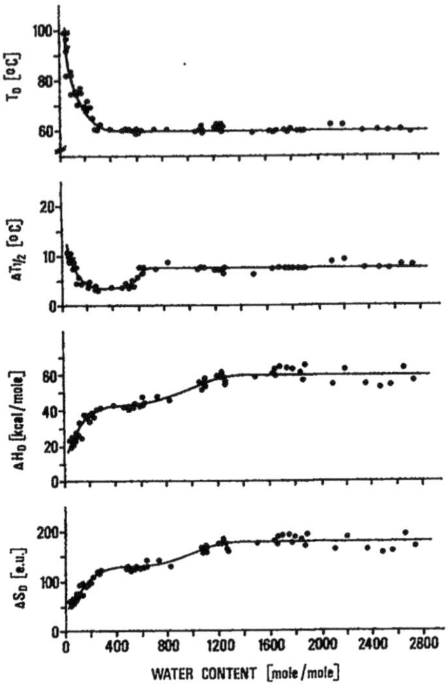

Fig. 4
Dependence on water content of thermodynamic parameters for thermal denaturation of α-lactalbumin.

ΔH_D: enthalpy of denaturation; T_D: temperature of denaturation; $\Delta T_{1/2}$: half width of the thermal transition; ΔS_D: entropy of denaturation, calculated form T_D and ΔH_D for $\Delta G = 0$ ($\Delta S_D = \Delta H_D/T_D$).

other proteins (22). This equation establishes a mathematical relation between the volume fraction of solvent in the polymer and the melting temperature of the polymer. In order to apply this type of equation one must know the transition temperature of the pure polymer. In the case of α-lactalbumin an extrapolation of the denaturation temperatures cannot be regarded as trustworthy because measurements at very low water contents (<30 mole/mole) are somewhat doubtful owing to the high temperatures of denaturation. Temperatures higher than 100°C may cause decomposition of the protein. If more than about 600 moles of water are absorbed the denaturation temperatures seem to remain constant within the experimental error.

The mean value of T_D and its standard deviation in the water content range 700-2800 mole/mole is $60.2 \pm 0.2^{\circ}C$.

The $\Delta T_{1/2}$ values, similar to those observed for β-lactoglobulin (3), decrease sharply upon absorption of primary hydration water to about $3.5^{\circ}C$. Absorption of secondary hydration water, however, has an opposite effect on the sharpness of the thermal transition. In the region of secondary hydration water the denaturation peak is broadened, i.e. $\Delta T_{1/2}$ increases in this region and reaches a plateau if more than about 800 moles of water are associated per mole of protein. The mean value for $\Delta T_{1/2}$ in the plateau region (800-2800 mole/mole) is $7.5 \pm 0.1^{\circ}C$.

Whereas the $\Delta T_{1/2}$ curve passes through a minimum in the region of secondary hydration water, a plateau is observed for the curves for the endothermic enthalpy and entropy changes. In the region of primary hydration water both functions, ΔH_D and ΔS_D, decrease sharply to values as low as 20 kcal/mole and 50 e.u. respectively at the lowest level of hydration tested. A second plateau is attained if approximately 1200 moles of water are associated with the protein. The mean values above this water content, in the range of 1200-2800 mole/mole, are 59.8 ± 1 kcal/mole for ΔH_D, and 179 ± 3 e.u. for ΔS_D.

Concentration Dependence of Thermal Denaturation

It was shown in Fig. 4 that the parameters of the thermal transition of α-lactalbumin, T_D, $\Delta T_{1/2}$, ΔH_D and consequently ΔS_D, remain practically constant in the water content range of about 1200 to about 3000 mole/mole. These water contents correspond to protein concentrations of 40 to 20 wt.%. Additional experiments were performed at lower protein concentrations in order to see if further dilution of the protein in water affects its heat stability. The results obtained are summarized in Table 1. Lowering of the protein concentration from 20% to about 10% in water shifts the mean transition temperature from 60.2 to $63.7^{\circ}C$ and broadens the denaturation peak. The enthalpy of denaturation, however, remains constant within the

Table 1

Heat denaturation of α-lactalbumin in concentrated and dilute solutions. Mean values for the temperature of denaturation (T_D), half width of the denaturation peak ($\Delta T_{1/2}$), calorimetric enthalpy of denaturation (ΔH_D) and calculated values of the entropy (ΔS_D) and van't Hoff enthalpy (ΔH_{vH}) of denaturation from DSC data.

Protein concentration (% w/w)	Solvent	T_D (^oC)	$\Delta T_{1/2}$ (^oC)	ΔH_D (kcal/mole)	a) ΔS_D (e.u.)	b) ΔH_{vH} (kcal/mole)	c) Renaturation (% of ΔH_D)
40 - 20	water	60.2 ± 0.2	7.5 ± 0.1	59.8 ± 1.0	179 ± 3	117	60 - 70
10 - 3	water	63.7 ± 0.4	10.7 ± 0.2	60.4 ± 1.2	179 ± 4	84	60 - 70
10 - 3	0.1 M phosphate buffer pH 6.9	64.8 ± 0.2	11.5 ± 0.3	65.1 ± 1.0	193 ± 3	79	80 - 90

a) Calculated from experimental ΔH_D and T_D values for $\Delta G = 0$: $\Delta S_D = \Delta H_D / T_D$.

b) Calculated from the mean T_D and $\Delta T_{1/2}$ values : $\Delta H_{vH} = 4RT_D^2 / \Delta T_{1/2}$ (21).

c) Estimated from DSC thermograms of samples which have been denatured, cooled at a rate of 5°/min and rescanned.

experimental error.

As judged from a second DSC scan of a sample which has previously been denatured, the rate or extent of renaturation was not increased by diluting the protein. 60-70% of the original area of the denaturation peak were obtained after cooling of the denatured sample at a rate of 5°/min.

The sharpness of a thermal transition observed in a thermogram may be used to evaluate the van't Hoff enthalpy (ΔH_{vH}) for the reaction involved (21). It can be shown that for the midpoint of the transition the following formula is applicable: $\Delta H_{vH} = 4RT_D^2/\Delta T_{1/2}$, where R is the gas constant, T_D the transition temperature and $\Delta T_{1/2}$ the half width of the heat absorption peak. However, this approximate formula, which is derived from the van't Hoff equation, is only strictly valid for a two-state process of the type $A \rightleftarrows B$. Comparison of the enthalpy change estimated from the sharpness of the transition (ΔH_{vH}) and the enthalpy change determined from the peak area (ΔH_D) has therefore been used as a criterion for a two-state case (1, 2). For transitions with a more complex stoichiometry than indicated by the equation $A \rightleftarrows B$, there will be a lack of correspondence between ΔH_{vH} and the true calorimetric heat change ΔH_D (23). Intermediate steps, for example, broaden the transition and lower ΔH_{vH}. For phase transition in which there is intermolecular cooperation, the van't Hoff enthalpy is much larger than the calorimetric enthalpy.

The thermal transition observed for α-lactalbumin in water does not meet the requirements for a two-state process. The ΔH_{vH} are greater than the corresponding calorimetric values (Table 1).

Heat Denaturation in Electrolyte Solution

Takesada et al. (5) when studying the heat stability of α-lactalbumin by ultraviolet absorption and circular dichroism measurement observed a thermal transition with a midpoint at 54°C and enthalpy and entropy changes of 72 kcal/mole and 220 e.u. respectively. In order to see if the discrepancy between these data and the data obtained by calorimetric methods was due to solvent effects, subsidiary experiments were carried out in phosphate buffer solutions. The results obtained are included in Fig. 3 and Table I. The change from pure water to the phosphate buffer solution increased the thermostability of α-lactalbumin and broadened the transition peak. Therefore, the van't Hoff heat decreased to about 79 kcal/mole, a value which may be compared to that observed by optical methods (72 kcal/mole (5)). However, ΔH_{vH} is still higher than the corresponding calorimetric heat, indicating that the process of denaturation can not be regarded as a simple two-state process which would involve only one initial and final state. The different temperature of denaturation reported by Takesada et al. (5) might be due to the difference in pH of the phosphate buffer solutions (5.7 compared to 6.9 in this study).

It is interesting to note that renaturation was also increased in the phosphate buffer system. Two factors might be responsible for the facility of reversible denaturation and the higher heat of denaturation in the presence of ionic substances: the capability of the ions to reduce the intermolecular electrostatic forces and the ion-specific structure stabilizing or destabilizing properties. Phosphate ions are known to have structure stabilizing effects since they appear in the Hofmeister series among the anions with strong salting-out and conformation stabilizing properties (24). A significant increase in the calorimetric enthalpy of denaturation upon addition of phosphate has also been observed for ribonuclease (25).

ACKNOWLEDGMENT

The authors wish to thank Dr. P. Lavanchy for amino acid analysis and Dr. M. Casey for linguistic assistance.

REFERENCES

1 Privalov, P.L.: Thermal investigations of biopolymer solutions and scanning microcalorimetry. FEBS Letters 40, S 140 - S 153 (1974).

2 Sturtevant, J.M.: Some applications of calorimetry in biochemistry and biology. Ann. Rev. Biophys. Bioengin. 3, 35 - 51 (1974).

3 Rüegg, M., Moor, U. and Blanc, B.: Hydration and thermal denaturation of β-lactoglobulin. A calorimetric study. Biochim. Biophys. Acta 400, 334 - 342 (1975).

4 Lüscher, M., Rüegg, M. and Schindler, P.: Effect of hydration on the thermal stability of tropocollagen and its dependence on the presence of neutral salts. Biopolymers 13, 2489 - 2503 (1974).

5 Takesada, H., Nakanishi, M. and Tsuboi, M.: Structure of α-lactalbumin and its fluctuation. J. Mol. Biol. 77, 605 - 614 (1973).

6 Lyster, R.L.J.: The denaturation of α-lactalbumin and β-lactoglobulin in heated milk. J. Dairy Res. 37, 233 - 243 (1970).

7 Bull, H.B. and Breese, K.: Thermal stability of proteins. Arch. Biochem. Biophys. 158, 681 - 686 (1973).

8 Townend, R. and Gyuricsek, D.M.: Heat denaturation of whey and model protein systems. J. Dairy Sci. 57, 1152 - 1158 (1974).

9 Armstrong, J.McD., Hooper, K.E., McKenzie, H.A. and Murphy, W.H.: On the column chromatography of bovine whey proteins. Biochim. Biophys. Acta 214, 419 - 426 (1970).

10 Brew, K., Vanaman, T.C. and Hill, R.L.: Comparison of the amino acid sequence of bovine α-lactalbumin and hens egg white lysozyme. J. Biol. Chem. 242, 3747 - 3749 (1967).

11 Quarfoth, G.J. and Jenness, R.: Isolation, composition and functional properties of α-lactalbumins from several species. Biochim. Biophys. Acta 379, 476 - 487 (1975).

12 Gál, S.: Recent advances in techniques for the determination of sorption isotherms, in Water Relations of Foods (Duckworth, R.B., ed.), pp. 139 - 154, Academic Press, London (1975).

13 Rüegg, M. and Blanc, B.: Effect of pH on water vapor sorption by caseins. J. Dairy Sci. 59, 1019 - 1024 (1976).

14 Kuntz, I.D., jr. and Kauzmann, W.: Hydration of proteins and polypeptides, in Advances in Protein Chemistry (Anfinsen, C.B., Edsall, J.T. and Richards, F.M., eds), Vol. 28, pp. 239-345, Academic Press, London (1974).

15 Lumry, R.: Participation of water in protein reactions. Ann. New York Acad. Sci. 227, 471 - 485 (1974).

16 Haly, A.R. and Snaith, J.W.: Calorimetry of rat tail tendon collagen before and after denaturation: The heat of fusion of its absorbed water. Biopolymers 10, 1681 - 1699 (1971).

17 Ramirez, J.E., Cavanaugh, J.R. and Purcell J.M.: NMR studies of frozen aqueous solutions. J. Phys. Chem. 78, 807 - 810 (1974).

18 Privalov, P.L. and Mrevlishvili, G.M.: Hydration of macromolecules in the native and denatured state. Biofizika 12, 22 - 29 (1967).

19 Rüegg, M., Lüscher, M. and Blanc, B.: Hydration of native and rennin coagulated caseins as determined by differential scanning calorimetry and gravimetric sorption measurements. J. Dairy Sci. 59, 387 - 393 (1974).

20 Haly, A.R. and Snaith, J.W.: Heats of fusion of various amounts of absorbed water in wool keratin. Biopolymers 7, 459 - 474 (1969).

21 Privalov, P.L., Khechinashvili, N.N. and Atanasov, B.P.: Thermodynamic analysis of thermal transitions in globular proteins. I. Calorimetric study of chymotrypsinogen, ribonuclease and myoglobin. Biopolymers 10, 1865 - 1890 (1971).

22 Flory, P.J. and Garett, R.R.: Phase transitions in collagen and gelatin systems. J. Am. Chem. Soc. 80, 4836 - 4845 (1958).

23 Brandts, J.F.: Conformational transitions of proteins in water and in aqueous mixtures, in Structure and Stability of Biological Macromolecules (Timasheff, S.N. and Fasman, G.D., eds.) 213 - 290, Marcel Dekker, Inc., New York (1969).

24 von Hippel, P.H. and Schleich, T.: The effects of neutral salts on the structure and conformational stability of macromolecules in solution, in Structure and Stability of Biological Macromolecules (Timasheff, S.N. and Fasman, G.D., eds.), 417 - 574, Marcel Dekker, Inc., New York (1969).

25 Tsong, T.Y., Hearn, R.P., Wrathall, D.P. and Sturtevant, J.M.: A calorimetric study of thermally induced conformational transitions of ribonuclease A and certain of its derivatives. Biochemistry 9, 2666 - 2677 (1970).

2.2 Comparative Studies of the Formation of Magnesium, Manganese and Cobalt Complexes with ATP Using Potentiometric and Calorimetric Techniques

M. Ragot, J.C. Sari, J. Galea, G. Ferroni, J.P. Belaich

The thermodynamic quantities associated with the formation of adenine nucleotide-magnesium complexes have been determined by many workers because of their biological interest. The almost totality of the kinases, particularly, needs ATPMg as substrate. However, many of these enzymes accept ATPMn and ATPCo complexes (1-7). A complete set of data (K_{app}, ΔH_{app}, ΔS_{app}) associated with Mg, Mn and CoATP complexes formation, has been obtained by an already described "pH-stat" titration method (8). The ΔH "Van t'Hoff" values are compared with ΔH "calorimetric" values measured in the same conditions.

EXPERIMENTAL SECTION

Commercial samples of sodium salt of ATP purchased from Boehringer, were submitted to further purification by ion exchange chromatography (DEAE Sephadex A-25).

The ionic strenght in all experiments was maintained at $\mu = 0.2$ with tetramethylammonium bromide for polarography (Merck) as the supporting electrolyte. During the experiments, the ionic strength variations do not exceed 5 % of μ and the activity coefficients were assumed constants. Triethanolamine has been used as buffer in all calorimetric experiments.

The "pH-stat" titration method is based on the lowering of

the pkp value of the second ionisation of the last phosphate group of the nucleotide when complexed with a divalent metal. The theory was published previously in detail (8).

Calorimetric ΔH determinations had been performed either with a LKB 10700-2 or a Tian-Calvet microcalorimeter. The enthalpy titration, used in these experiments, has been reported by Belaich et coll. (9) and Sari et coll. (10).

PRINCIPLE OF CALCULATION

We have checked that for a pH high enough (pH > 8) the following equilibrium was the only one to be considered :

$$ATP^{4-} + M^{2+} \rightleftharpoons ATPM^{2-}$$

If a and b_1 are the initial concentrations of ATP and metal ion respectively, and x_1 the concentration of complex at equilibrium, the apparent formation constant is :

$$(1) \quad K_{app} = \frac{x_1}{(a - x_1)\,(b_1 - x_1)}$$

When this reaction occurs at constant temperature and pressure in a calorimeter of utilizable volume (V), the heat evolved during the reaction is :

$$q_1 = x_1 . \Delta H_{app} . V$$

The heat evolved per liter will be :

$$(2) \quad Q_1 = x_1 . \Delta H_{app}$$

where ΔH_{app} is the apparent enthalpy of the complex formation reaction for $\mu = 0.2$. If a second experiment is done under the same conditions, but with a different initial concentration (b_2) of metal ion, the concentration of complex at equilibrium will be x_2, and :

$$(3) \quad K_{app} = \frac{x_2}{(a - x_2)\,(b_2 - x_2)}$$

and

(4) $Q_2 = x_2 \, \Delta H_{app}$

From relationships 1, 2, 3 and 4, we obtain :

$$(5) \quad \Delta H_{app} = \frac{(b_2 - b_1)\, Q_1\, Q_2 \pm \sqrt{\Delta}}{2\,(Q_1\, ab_2 - Q_2\, ab_1)}$$

with

$$\Delta = Q_1^2\, Q_2^2\, (b_2 - b_1)^2 - 4(Q_1\, ab_2 - Q_2\, ab_1)\, (Q_1\, Q_2^2 - Q_1^2\, Q_2)$$

Experimental calorimetric determination of Q_1 and Q_2 gives an estimation of ΔH_{app}, and the knowledge of ΔH_{app} provides K_{app} from relationships (1) and (2) or (3) and (4).

When the apparent association constant value is high enough ($> 10^5\ M^{-1}$) and ATP concentration about 10^{-4} M, the equilibrium

$$ATP^{4-} + M^{2+} \underset{}{\overset{K_{app}}{\rightleftharpoons}} ATPM^{2-}$$

can be entirely displaced towards complex formation by a rather small excess of metal. ΔH_{app} will be calculated by the relationship :

(6) $\Delta H_{app} = \frac{Q}{C}$

where C is the final nucleotide concentration, therefore the complex concentration.

If ATP is not entirely dissociated (for pH near 7), the equilibria are :

$$\begin{array}{l} ATP^{4-} + M^{2+} \overset{K_{app}}{\rightleftharpoons} ATPM^{2-} \\ H^{+} \Big\Updownarrow kp = 10^{-6.96} \quad (8) \\ ATPH^{3-} \end{array}$$

$ATPHM^{-}$ complex is not considered at this pH because his ionisation constant (kp) is higher than 10^{-5}.

The values of ΔH_{app} measured in this case (which we call $\Delta H_{app.meas.}$) must be corrected according to the ΔH of ionisation of nucleotide and buffer (ΔH_N and ΔH_T, respectively), and ΔH_{app} becomes :

$$(7) \quad \Delta H_{app} = \Delta H_{app.meas.} - \frac{(H^+)}{kp + (H^+)} \cdot (\Delta H_N - \Delta H_T)$$

RESULTS

Table I (a, b, c) shows the potentiometric results obtained for $ATPMg^{2-}$, $ATPMn^{2-}$ and $ATPCo^{2-}$ complexes, respectively

TABLE I - Potentiometric results

a) Complex $ATPMg^{2-}$, pH 8.5 (ATP) = 5.10^{-4} M

T° K	$ATPMg^{2-}$		
	K_{app} M^{-1}	ΔH_{app} kcal/mole	ΔS_{app} u.e.
276	21,700	4.50	36.1
290	34,100	"	"
299	41,900	"	"
303	45,700	"	"
312	55,800	"	"

b) Complex $ATPMn^{2-}$, pH 7.5 (ATP) = 3.10^{-4} M

T° K	$ATPMn^{2-}$		
	K_{app} M^{-1}	ΔH_{app} kcal/mole	ΔS_{app} u.e.
274	278,000	- 0.4	23.4
278	279,000	1.6	30.7
283.5	300,000	4.1	39.5
288	350,000	5.9	45.8
298	510,000	9.0	56.3
308	910,000	11.3	64.0
316	1,500,000	12.6	68.1

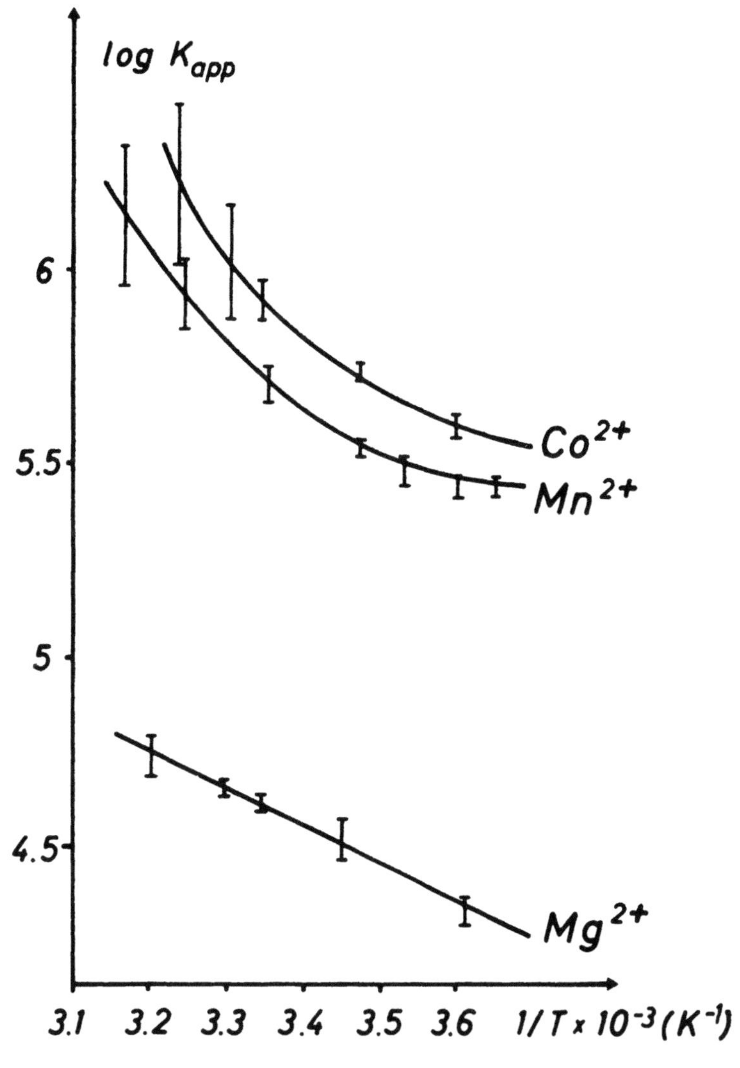

Figure 1 - Van t'Hoff isochores for different $ATPM^{2-}$ complexes

c) Complex $ATPCo^{2-}$, pH 7.1 (ATP) = 10^{-4} M

T° K	$ATPCo^{2-}$		
	K_{app} M^{-1}	ΔH_{app} kcal/mole	ΔS_{app} u.e.
277.8	380,000	3.4	37.8
287.8	540,000	5.9	46.9
299	820,000	10.5	62.4
303.4	1,250,000	12.5	69.5
309	1,740,000	16.4	81.8

Van t'Hoff isochore for each complex studied is represented on figure 1. For $ATPMg^{2-}$, we obtain a straight line thus ΔH_{app} is constant and equal to 4.50 $\pm$ 1.34 kcal/mole with a confidence level of 0.95. In the case of $ATPMn^{2-}$ and $ATPCo^{2-}$ complexes, a non linear temperature dependence is observed. The ΔH_{app} of the formation reaction of these two complexes vary in a large interval (figure 2). It has been determined by polynomial regression, so it is not possible to calculate confidence intervals.

Figure 3 and Table II show the calorimetric results obtained for the three complexes studied. The metal does not seem to have any influence on the ΔH_{app} of the complex-reaction formation. We can explain the decrease of the ΔH_{app} values, for $ATPMn^{2-}$ complex, with the temperature by an association of the nucleotide molecules induced by manganese ions. Indeed, Gill et al. (11) had observed that the self-association of purine and purine ribosides is an exothermic reaction.

From the potentiometric and calorimetric data we can compare ΔH van t'Hoff and ΔH calorimetric values obtained in the same experimental conditions. In the case of $ATPCo^{2-}$ and $ATPMn^{2-}$

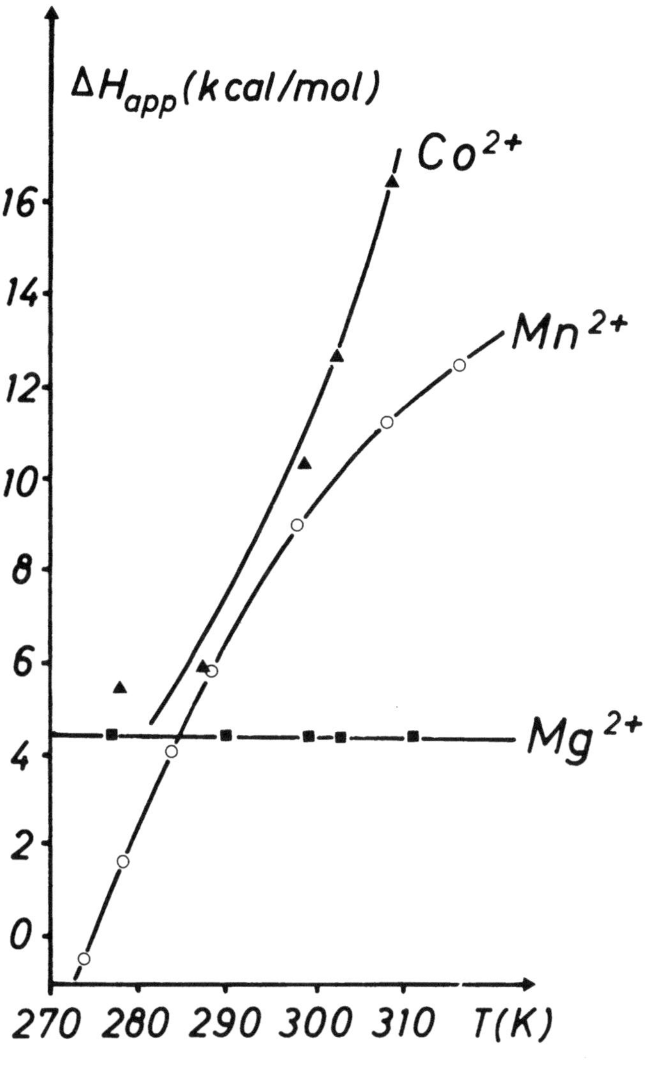

Figure 2 - Van t'Hoff ΔH values versus temperature for different $ATPM^{2-}$ complexes.

TABLE II - Calorimetric results

ΔH_{app} Kcal/mole	T° C			
	30	25	15	6
$ATPMg^{2-}$	4.48±0.05	-	-	-
$ATPMn^{2-}$	4.63±0.15	-	3.66±0.25	3.17±0.11
$ATPCo^{2-}$	-	4.55±0.15	-	-

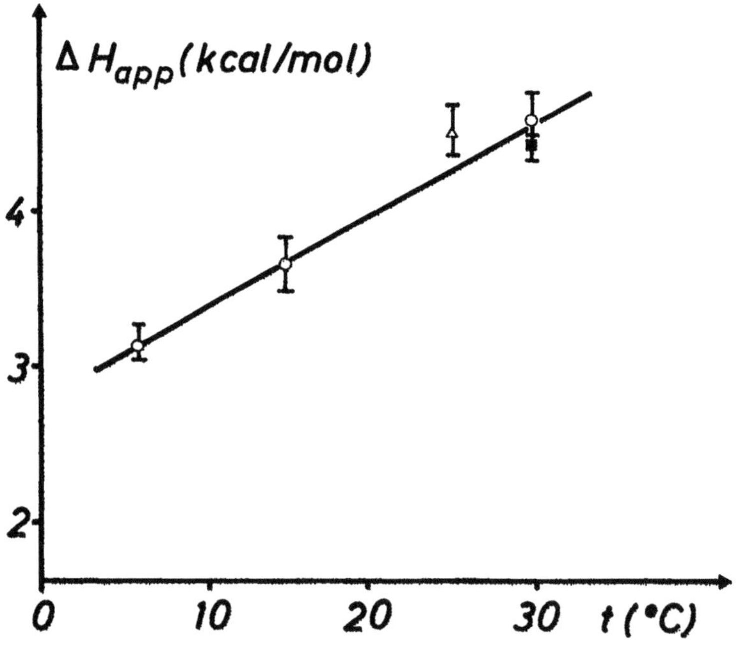

Figure 3 - Calorimetric ΔH values versus temperature. o $ATPMn^{2-}$; Δ $ATPCo^{2-}$; ■ $ATPMg^{2-}$.

the K_{app} obtained by the potentiometric technique are too high and unprecise to calculate accurate ΔH values, so calorimetry is the only method which can be used for this determination.

On the other hand, for $ATPMg^{2-}$ whose K_{app} values are weaker, the two ΔH are in good agreement, but the standard error on Van t'Hoff ΔH is greater than that obtained by calorimetry (4.50 ± 1.34 kcal/mole and 4.48 ± 0.05 kcal/mole respectively).

REFERENCES

1. Larsson-Ràznikiewicz, M. and Malmström, B.G.: The metal ion activation of 3-phosphoglycerate kinase in correlation with metal-binding studies. Arch. Biochem. Biophys., 92, 94-99 (1961).
2. Cohn, M.: Magnetic resonance studies of metal activation of enzymatic reactions of nucleotides and other phosphate substrates. Biochemistry, 2, 623-629 (1963).
3. O'Sullivan, W.J. and Cohn, M.: Magnetic resonance investigations of the metal complexes formed in the manganese activated creatine kinase reaction. J. Biol. Chem., 241, 3104-3115 (1966).
4. Mildvan, A.S. and Cohn, M.: Kinetic and magnetic resonance studies of the pyruvate kinase reaction. II. Complexes of enzyme, metal and substrates. J. Biol. Chem., 241, 1178-1194 (1966).
5. Cazzulo, J.J. and Stoppani, A.O.M.: Effects of magnesium, manganese and adenosine triphosphate ions on pyruvate carboxylase from Baker's yeast. Biochem. J., 112, 747-754 (1969).
6. Reed, G.H. and Cohn, M.: Analysis of equilibrium data from proton magnetic relaxation rates of water for manganese-nucleotide-kinase ternary complexes. J. Biol. Chem., 245, 6547-6552 (1970).
7. Reed, G.H. and Cohn, M.: Electron paramagnetic resonance studies of manganese. (II)-Pyruvate kinase-substrate complexes. J. Biol. Chem., 248, 6436-6442 (1973).

8. Sari, J.C., Ragot, M. and Belaich, J.P.: "pH-Stat" titration method for the determination of the thermodynamic quantities associated with the formation of ATP-magnesium complexes. Biochim. Biophys. Acta, 305, 1-10 (1973).
9. Belaich, J.P. and Sari, J.C.: Microcalorimetric studies of the formation of magnesium complexes of adenine nucleotides. Proc. Nat. Acad. Sc. USA, 64, 763-770 (1969).
10. Sari, J.C. and Belaich, J.P.: Microcalorimetric studies on the formation of magnesium complexes with 5' ribonucleotides of guanine, uracil and hypoxanthine. J. Am. Chem. Soc. 95, 7491-7496 (1973).
11. Gill, S.J., Downing, M. and Sheats, G.F.: The enthalpy of self-association of purine derivatives in water. Biochemistry, 6, 272-276 (1967).

2.3 Microcalorimetric Experiments on Cell Free-Protein Biosynthesis

L. Berthe-Corti

Heat production associated with protein biosynthesis (PBS) in different cell-free systems of yeast (Saccharomyces) and bacteria (E.coli) was investigated.

The following steps were elaborated:

1. realisation of the complete synthesis (with all essential components)
2. inhibition of PBS (above all antibiotics) to distinguish heat produced by PBS and heat produced by interfering reactions
3. hydrolysis of RNA,ATP and GTP related to PBS

1. Realisation of the Complete Synthesis

The methods of isolation of the cell fractions as well as the biochemical assays are described elsewhere (1,2), as are the microcalorimetric investigations (1). The latter were made with a microcalorimeter of the type E. Calvet.
The course of PBS is determined by a sum of different factors and enzymes that interact directly in the process, such as the enzymes ATPase and GTPase (3,4).
Apart from this, there are many associated reactions, reactions which supply substrates and reactions, which are regulatory mechanisms for example ribonuclease activity (5,6). With our method of isolating the cell fractions, it was not possible to isolate all these factors separately. Therefore it was necessary to optimise the systems as to their biochemical behaviour. The optimised systems were tested in parallel for their biochemical and microcalorimetrical behaviour.
As a first aproximation to the heat production of PBS in complete systems, the following tests were made:

a. The incorporation of ^{14}C-phenylalanine (spec. activity 50 μC/mol) in different complete systems was investigated.
Parallel to this, the heat-production of different complete systems was tested as is to be seen in figure 1.

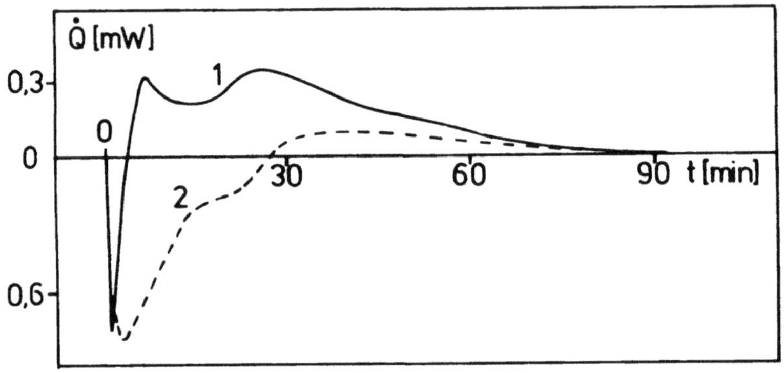

FIG. 1 Thermograms of complete systems: cell-free protein synthesizing systems of yeast and bacteria.
1: Yeast complete system (spec. incorporation of ^{14}C-phenylalanine 5900 cpm)
2: Bacterial complete system according to Nomura and Erdmann (7) (spec. incorporation 78000 cpm)

The course of the curves is typical for eucaryotic and procaryotic systems, but not exactly reproducible.

b. The dependence of the heat production on essential cell components was tested. As an example, the dependence on the enzyme fraction (EF) is shown. The dependence is very small except for the components ribosomal fraction (RF) and MIX (MIX, see caption under figure 3) as can be seen in figure 2.

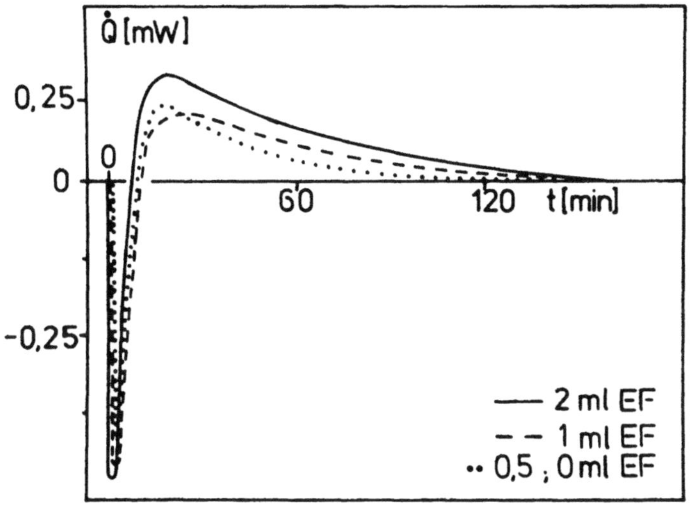

FIG.2 Course of thermograms of a yeast complete system as a function of the amount of the enzyme fraction (EF).

However in the biochemical assay, we found a relatively good dependence of the peptide formation on the amount of enzyme. The incorporation of ^{14}C-phenylalanine into the peptide was reduced by 50-70% when the enzyme fraction was ommitted compared with the complete system.

c. The dependence of the incorporation of the marked amino acids on the linkages rich in energy (that is to say MIX) and the dependence of the heat production on MIX do not follow the same course. Figure 3 shows that the heat production is proportional to the amount of MIX. Here this is shown with two independent isolations of yeast. Figure 3 shows also that we found a clear optimum in the biochemical assay.

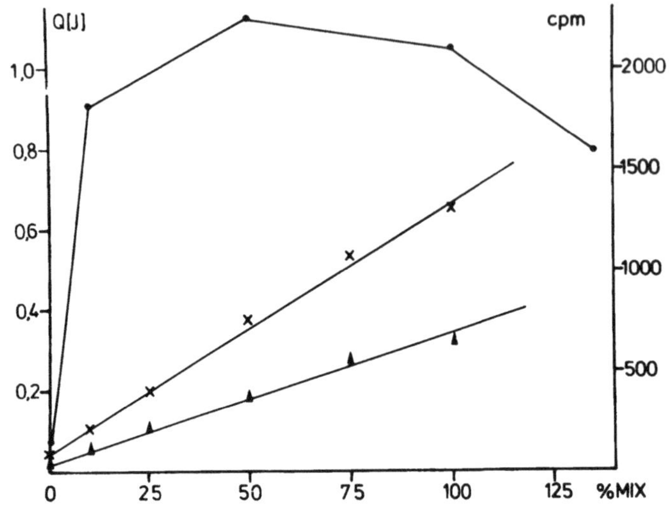

FIG.3 Heat production of two different complete systems of yeast as a function of MIX. Thermograms and incorporation data of ^{14}C phenylalanine into the polypeptide.

MIX: 7.5 mg/ml ATP; 7.5 mg/ml GTP, 12.5 mg/ml PeP, 25 μl PeP-kinase, pH 7.5. 100% MIX: in the biochemical assay = 50 μl, in the microcalorimetric assay = 500 μl.

d. A partial reaction was tested by incorporation of ^{14}C phenylalanine into the peptide. The assay included only the amino acids and the ribosomes in a magnesium buffer ($1 \cdot 10^{-2}$M Mg^{++}). In the biochemical tests we did not observe peptide formation . However with the microcalorimeter we found that a mixture of ribosomes and amino acids reveal an exothermal heat production.

The above-mentioned results support the assumption that with the methods of isolation used, the ribosomal fraction in particular is not free of interfering reactions.

Therefore the suggestion was made, that in the microcalorimeter we had also measured effects that sometimes do not have a noticeable influence on PBS in the biochemical assays. Two such reactions were considered:

- Changes in the configuration of molecules, especially those of the ribosomes, and
- chemical reactions, as for example enzymatic processes.

A rough evaluation with data given by Traub (8) for the configuration enthalpy of ribosomes showed that the above effects cannot be produced by changes in the configuration or by dissociation of ribosomes.
That means that the problem is to distinguish heats that are produced by PBS itself and those that are caused by interfering reactions.

2. Experiments on Inhibition of PBS to Distinguish Heats, that are the Result of PBS, and those that Arise from Interfering Reactions

For this, a simple method seemed to be the investigation of specific inhibition of PBS. Inhibitors (above all antibiotics) that block specific steps of PBS were used.

TABLE 1 Incorporation of ^{14}C- Phenylalanine into the Complete System of Yeast and connected Heat Production

conditions	inhibitorconc. and polypeptide formation in % of the complete system					heat production (m J)	
	data by the author		data by other authors			Q^{+} (+)	Q^{-} (++)
complete system		100		100		404 ± 84	361 ± 84
Cyclohexi-mide	1 mM	22 ± 8	1,0 mM 0,01 mM	15-30 14	(9) (10)	322 ± 84	27 ± 29
Puromycin	1 mM	29 ± 10	0,1 mM 0,1 mM	0 38	(11) (10)	94 ± 42	21 ± 20
Fusidic acid	1 mM	17 ± 10	1,0 mM	7	(11)	no significant difference to complete system	

(+) endothermal part of the thermogram
(++) exothermal part of the thermogram

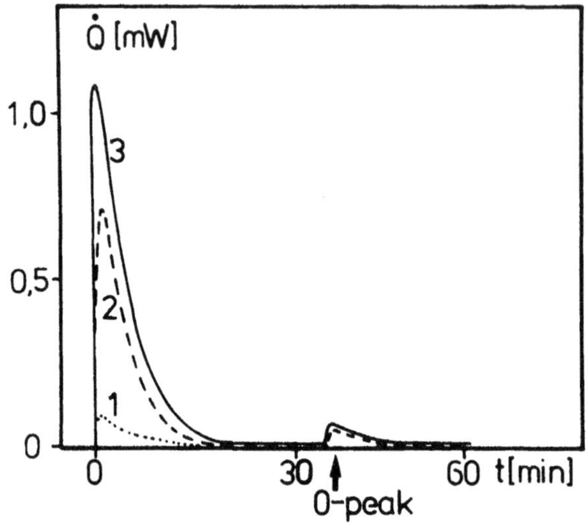

FIG.4 Thermograms of the RNA degradation by the ribosomal fraction (isolated from yeast) Assay: 2.5 ml RNA (12 mg/ml); 3.0 ml TRIS-HCl buffer pH 7.5, ribosomal concentration OD_{260}= 400 curve 1:1 ml RF, 2: 0.5 ml RF, 3: 0 ml RF

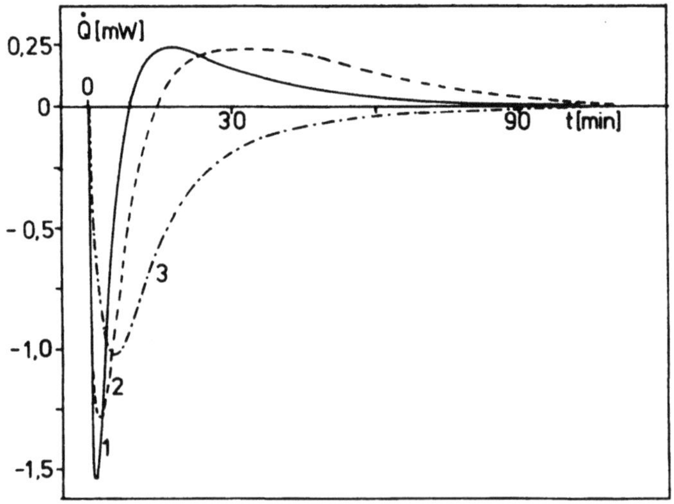

FIG.5 Course of thermograms of RNA degradation as a function of the amount of commercial ribonuclease. Assay: 2.5 ml RNA (30 mg/ml); 1,5 ml TRIS-HCl,EDTA-buffer pH 7.5; 1,0 ml ribonuclease. Ribonucl.conc: curve 1: 1.5 mg/ml; 2:0.15 mg/ml; 3: 0.015 mg/ml.

In the biochemical test, inhibition values that were anticipated by the literature were found. The results obtained are shown in table 1. The results of the microcalorimetric investigations reveal that the exothermal part of the thermograms is somewhat diminished compared with that of the complete system. However, the course of the thermograms allow no exact elucidation of particular steps of PBS. The method of elucidating complete PBS system with the aid of specific inhibitors thus did not lead to a satisfying result. This is not only a problem of the complexity of the different biochemical reactions, but also a problem of the method of measurement. The instrument used shows an integral signal from all simultaneously occuring processes. This suggests the use of single processes associated with PBS for the discrimination of the various sources of heat production and for getting an aproximation to the complete heat.

3. Enzymatic Reactions that are Essentially Connected to PBS

For this procedure three reactions were chosen:
a) the hydrolysis of RNA by ribonuclease and
b) and c) the hydrolysis of ATP by ATPase and GTP by GTPase.

a) The hydrolysis of RNA by ribonuclease. This reaction is important because we did not suceed in isolating active fractions free of ribonuclease activity. The ribonuclease activity is bound to the ribosomes and has a central position in the regulation of PBS (12, 13). In the biochemical assay, degradation runs spontanuously without consumption of linkages rich in energy such as ATP or GTP. As is shown in figure 4, the thermograms reveal an exothermal course. The heat production is a function of the amount of the ribosomal fraction. In addition to the ribosomal fraction we tested the activity of commercial ribonuclease. With a combination of RNA and commercial ribonuclease, the endothermal part of the thermograms predominates. In figure 5, thermograms with constant RNA concentration and decreasing amounts of ribonuclease are shown. The thermograms reveal similar courses with various concentrations of RNA.

In this way we found that the heat production was positive only with a combination of RNA and ribosomes. That means that the heat production in assays containing ribosomes is not only due to the hydrolysis of RNA.

b) <u>hydrolysis of ATP by ATPase</u> Some authors found ATPase and GTPase activity bound to the ribosomes (14,3). H. Lother succeded in demonstrating that ATPase and GTPase activities are bound to yeast ribosomes (15). Microcalorimetric investigations were made to reveal the heat production that is combined to this activity. Figure 6 shows the course of the thermograms of the ATP hydrolysis on the ribosomal- and the ATP-concentrations. The course of the thermograms is not reproducible.

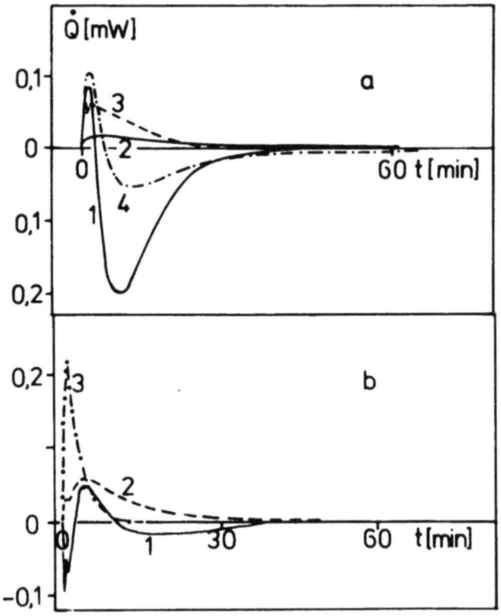

Fig.6 Thermograms of the ATP hydrolysis (different assays with different ATP or ribosome concentrations, ribosomes isolated from yeast).

a) varation of ribosome conc. (ATP conc.: $1\cdot 10^{-3}$M)

1: OD_{260}=9; 2: OD_{260}= 45; 3: OD_{260}= 90; 4: OD_{260}= 180;
5: OD_{260}= 0 (thermogram 5 = 0-line)

b) ribosome conc.: OD_{260}= 90
variation of ATP conc. of the curves: 1: $2\cdot10^{-3}$ M ATP;
2: $5\cdot10^{-3}$ M ATP; 3: $1\cdot10^{-2}$ M ATP.

assay: 2.7 ml ATP solution (buffer MgCl : from $1\cdot10^{-3}$ M to $1\cdot10^{-2}$ M; KCl: $1\cdot10^{-1}$ M; TRIS-HCl: $1\cdot10^{-1}$ M; Mercaptoethanol: $6\cdot10^{-3}$ M; pH 7.8) and 0.3 ml ribosomes

c. <u>hydrolysis of GTP by GTPase</u> The mode of treatement of the GTPase reaction ran parallel to the ATPase reaction. Figure 7 shows the course of the thermograms with different amounts of ribosomes.
In both cases (ATPase and GTPase reaction) a behaviour was observed in which the heat production is far below the one anticipated by the hydrolysis energy.

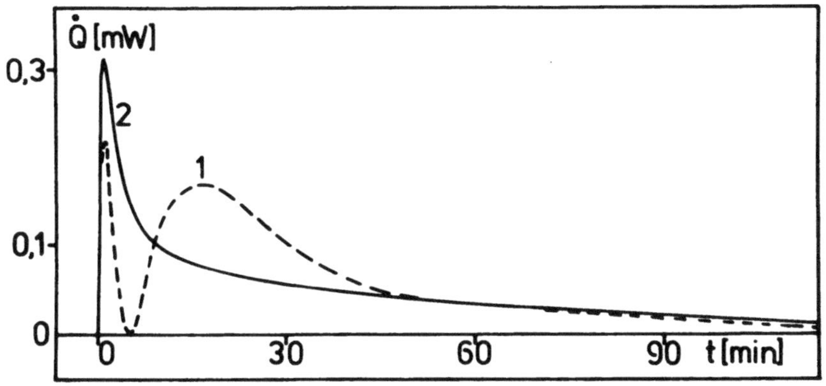

FIG.7 Thermograms of the GTP hydrolysis by ribosomes isolated from yeast.
assay: 2.0 ml GTP solution (buffer A-P; pH 7.5)
GTP conc. 3 mg/ml, 0.5 ml ribosomes
ribosome conc.: curve 1: OD_{260}=600, curve 2: OD_{260}=900

The facts that in all systems :
1. the omission of essential components other than RF or MIX has almost no influence on the thermograms,
2. the effect of inhibition of PBS by antibiotics shown in the biochemical assay could not be shown in the microcalorimeter,

3. a mixture of ribosomes and MIX gave an exothermal heat production without the possibility of PBS occuring,
4. GTPase and ATPase activity showed endo- and exothermal heat production,
5. ribonuclease activity with ribosomes always revealed an exothermal heat production, whereas RNA degradation with commercial ribonuclease showed thermograms with endo- and exothermal parts,

show that the thermograms obtained are related to PBS but not exclusivly due to PBS.

The problems showed above make it clear that the heat fluxes in the microcalorimeter that are measured during the protein biosynthesis cannot be directly traced back to PBS.

These investigations were performed at the Freie Universität Berlin, Zentralinstitut für Biochemie und Biophysik,

The excellent technical assistance of Mrs. G.Welge is gratefully acknowledged.

REFERENCES

1. Berthe-Corti,L., Mikrokalorimetrische Untersuchungen zur Proteinbiosynthese in zellfreien Systemen aus Hefe, Dissertation, Freie Universität Berlin, (1975).

2. Schulz-Harder,B., Untersuchungen zur zellfreien Proteinbiosynthese in einem Saccharomyces-System, Diplomarbeit, Freie Universität Berlin, (1973).

3. Grummt,F., Grummt,I. and Erdmann,V.A, ATPase and GTPase Activities Isolated from Rat Liver Ribosomes, Eur.J. Biochem. 43, 343 - 348 (1974).

4. The Mechanism of Proteinbiosynthesis and its Regulation, ed .L.Bosch, North Holland Publ. Comp. (1973).

5. Lucas-Lenard,J. and Lipmann,F., Proteinbiosynthesis, Ann.Rev. Biochem. 40, 409 - 438 (1971).

6. Kössel,H. and Bhandary,U.L.R., Studies on Polynucleotides, J.Mol.Biol. 35, 539 - 560 (1968).

7. Nomura,M. and Erdmann,V.A., Reconstitution of 50-S-ribosomal Subunits from dissociated molecular components, Nature 228, 744 - 748 (1970).

8. Traub,P. ,Current Topics of Microbiology and Immunology, Vol.52,P 1-93 Springer Verlag Berlin (1970).

9. Vazquez, D., Battaner,E., Neth,R., Heller,G.,and Monro,R., The Function of 80-S-ribosomal Subunits and Effects of some Antibiotics, Cold Spring Harbor Symp. Quant.Biol. 34, 369 - 375 (1969).

10. Chester,J.K., Proteinsynthesis by Cell-Free Extracts of Crithidia oncopeeti, Biochim.biophys.Acta 114, 385 - 397 (1966).

11. Battaner,E. and Vazquez,D., Inhibitors of Proteinsynthesis by Ribosomes of the 80-S-Type, Biochim.biophys.Acta 254, 316 - 330 (1971).

12. Chet,I., Rettig,N. and Henis,Y., Changes in Ribonuclease during Differientiation (sperulation) of Physarum Polycephalum Biochim. biophys.Acta 294, 343 - 347 (1973)

13. Ceri,H. and Maeba,P.Y., Association of a Ribonuclease with the 50-S-ribosomal Subunit of E.Coli MRE 600, Biochim.biophys.Acta 312, 337 - 348 (1973)

14. Horne,J. and Erdmann,V.A., ATPase and GTPase Activities associated with a specific 5-S-RNA-Protein Complex, Proc.Nat.Acad.Sci. (USA) 70, 2870 - 2873 (1973)

15. Lother,H., unpublished

2.4 Enthalpy Changes in Reactions of Messenger-RNA Turnover

L. Tumerman, S. Ric

The high rate of turnover is a known peculiarity of the m-RNA. In some procariotic cells like those of E.coli and B.subtilis, the mean chemical lifetime of a m-RNA molecule has been found to be in the range of 2 to 6 minutes (1-7). The transcription of the m-RNA represents 50 to 80% of the overall RNA transcription in these cells, though the m-RNA is the least abundant among all RNA's and makes up only a small percent of the total cell RNA.

In eucariotic cells the mean lifetime of m-RNA molecules may be different for cell derived from various tissues, but in general it is greater than in procariotic cells. However, only a small part (approx. 10%) of the long RNA chain transcribed on a cistron of these eucariotic cells represents the "true" messenger-RNA which is transferred to the ribosomes and acts as a template in protein biosynthesis. The rest is decomposed into monomeric units immediately after the transcription has been finished and does not leave the cell nucleus (8). The rate of the m-RNA turnover is therefore very high in these cells too.

The informational, or functional, meaning of the high rate of m-RNA turnover is easily understandable, but the energetics of this process is far from being so clear. The energy of two macroergs is required to rephosphorylate the monomeric units of the RNA to the level of precursors. Now the question arises: Does this energy come from the cell's general energetic resources, i.e. from its pool of ATP molecules, or is the energy released by the breakdown of macroergic bonds in the precursor stored by some way in the system and used later to perform the required phosphorylations? In other words: Is the cycle of reactions involved in the m-RNA turnover an "open" or a "closed" one?

Experimental

Reactions Studied

Heat effects were measured and the corresponding values of enthalpy change (ΔH) were computed for the following enzymatic reactions:

1. Hydrolysis of ADP catalyzed by the Apyrase from potatoes

$$ADP \xrightarrow{Apyrase} AMP+P_i$$

2. Formation of poly-A from ADP molecules catalyzed by the Polynucleotide Phosphorylase (CE 2.7.7.8)

$$n(ADP) \xrightarrow{PNP'se} [poly\text{-}A]_n + n\ P_i$$

3. Phosphorolysis of poly-A to ADP molecules catalyzed by the same enzyme

$$[poly\text{-}A]_n + nP_i \longrightarrow n(ADP)$$

4. Hydrolysis of poly-A and RNA to single nucleotide monophosphate units catalyzed by the Ribonuclease A5 from Actinomyces sp.

$$[poly\text{-}A]_n \longrightarrow n\ (AMP)$$

The enzyme was isolated and studied by Tatarskaya et al. (9).

Reaction media contained as follows:

For the reaction I: Succinate buffer 0.04M, pH 6.8 + $CaCl_2$ 0.004M;
For reactions 2 and 3: Tricine buffer 0.1M, pH 8.2 + $MgCl_2$ 0.005M.
All reactions were performed at 25-27°C.

Heat Measurements

The heat effects of the above mentioned reactions were measured with the LKB batch microcalorimeter. The readings of this instrument, which belongs to the type of calorimeters based on the heat conduction (or heat leakage) principle, are proportional to the heat power of the reaction, i.e. to the rate of heat evolution or absorption. The amount of heat evolved or absorbed in the course of the reaction was computed from the area under the registered voltage-time curve using the appropriate calibration of the instrument.

The substrate solution and appropriate buffer mixture, brought to a total volume of 4ml, were poured into the greater compartment of both the reaction and the reference cell of the instrument. The smaller compartments of both cells were filled with 2ml of buffer solution without the substrate. Then 20 to 50µl of enzyme solution were poured into the smaller compartment of the reaction cell only. After thermal equilibration of the instrument the reaction was initiated by rotating the calorimeter and mixing the liquids in both cells. It should be noted that all heat effects due to the dilution of the substrate solution, to small changes in pH etc. were eliminated except for the effect of the enzyme dilution which had been found to be negligibly small.

To compute the value of the enthalpy change (ΔH) characteristic for the reaction under investigation, one only has to divide the measured heat effect $\underset{\sim}{Q}$ by the number $\underset{\sim}{N}$ of bonds disrupted or newly formed in the course of the reaction.

All measurements were repeated 5 to 10 times and the results were averaged. The mean quadratic error had been found to be in the range of 10 to 20% of the measured ΔH value.

Analytical Methods

In reactions 1 and 2 the number $\underset{\sim}{N}$ of bonds disrupted and/or newly formed was established by measuring the concentration change of free orthophosphate. In reactions 3 and 4 it was established by measuring concentration changes of corresponding nucleotide phosphates.

Orthophosphate concentration was measured by the way of a colour reaction with molybdate, H_2SO_4 and a reducing agent. The reagent consisted of 1 part of Vit. C 10% solution, 6 parts of 0.42% NH_4 Mo solution in 1M H_2SO_4 and 3.5 parts of water. 10µl of the specimen were added to 1ml of the reagent, and the mixture was incubated for 30 minutes at 45^oC. After cooling to the room temperature the optical density of the solution was measured at 820nm. In calibration experiments, performed with KH_2PO_4 solutions of known concentration, strict proportionality has been established between the concentration of the orthophosphate and the measured values of OD_{820} up to at least OD_{820}=1,2.

AMP and ADP concentrations were measured by the thin-layer chromatography on PEI-cellulose (Merck 5666, TLC aluminium sheets). 5 to 10µl of the specimen and of appropriate markers (AMP, ADP and ATP solutions of known concentration) were applied to the start-line of the sheet, and chromatograms were developed with 0.2M LiCl for 10 minutes, then with 1.2M LiCl for 20 minutes and finally by 2.0M LiCl. The nucleotide spots were examined and identified under a short-wave ultraviolet lamp and then eluted with 1ml of the mixture: Tris buffer 1.5M, pH 7.4 + 0.75M $MgCl_2$. Optical density of the eluate was measured at 260nm.

Chemicals

Nucleotide phosphates and enzymes were purchased from "Sigma" and used without any further purification, except for the Ribonuclease A5 which was a gift from Dr. R. Tatarskaya (Institute of Molecular Biology, Moscow). Some experiments were also performed with the solution of the Polynucleotide Phosphorylase isolated from E.coli and purified by Dr. M. Shoreq (The Weizmann Institute of Science, Rehovot, Israel). No essential difference in ΔH-values were found between these two preparations of the enzyme.

Results and Discussion

The results of our measurements are presented in Table 1.

Table 1 - ΔH-values for the reactions studied

Reaction	ΔH (kcal/mol)
ADP Hydrolysis	-11.2
Poly-A Synthesis	- 1.7
Poly-A Phosphorolysis	+ 3.3
Poly-A (and RNA) Hydrolysis	- 9.5

The value ΔH=- 11.2 kcal/mol that we have found for the reaction of ADP hydrolysis to AMP and P_i is close to the value - 12 kcal/mol found previously (10) for the ATP hydrolysis to ADP and P_i.

Thermodynamically the process of poly-A (or RNA) formation represents the coupling of the exothermic reaction of the precursor (ADP or NTP) hydrolysis and the endothermic reaction of NMP polymerization. The comparison of ΔH-values for the first two reactions in the Table shows that 9 kcal/mol or approximately 85% of the energy liberated by the breakdown of the macroergic bonds in the precursor, are not released as heat but rather are stored in the newly formed 3',5'-phosphodiester bridges. Our data for the reactions 3 and 4 show that this stored energy can either be released as heat in the reaction of hydrolysis, or used to re-phosphorylate the nucleotide monophosphates in the phosphorolytic reaction, which represents the coupling of the RNA hydrolysis with the AMP phosphorylation. In the model cycle of RNA turnover, which consists of RNA synthesis from ADP and the reverse of this reaction, the energy is transferred from (P-O-P)-bonds to (P-O-P) bonds and back. The transfer proceeds with a very high efficiency.

The pattern of the reaction cycle for the natural process of m-RNA turnover is more complicated. It has been shown (6) that in vivo the breakdown of m-RNA does not proceed to the level of nucleotide monophosphates. The degradation products enter the nucleotide diphosphate pool directly and not via the nucleotide monophosphates. But since only the nucleotide tri-phosphates can act as precursors in the natural process of RNA transcription, and pyrophosphate groups, instead of the orthophosphate, are liberated in this process, the nucleotide diphosphates must be phosphorylated to close the cycle. This cycle therefore consists of at least three steps, as shown in Fig. 1.

The first two steps of the cycle resemble, from the energetic point of view, the model cycle that we have investigated. Since at P_i concentrations usually present in cells the reaction catalyzed by the PNP's proceeds in the direction of RNA phosphorolysis, it seems natural to assume that the phosphorolytic breakdown of m-RNA is the biological function of this enzyme.

L. Peller (11) has shown that without the hydrolysis of liberated pyrophosphates, thermodynamics limits the number average degree of polymerization to less than 100 nucleotides in size. To explain the formation of much longer chains of RNA transcribed on a cistron, one must admit that RNA transcription is coupled to the pyrophosphates hydrolysis. That raises the upper limit of the average size of the polymer by a factor of the order of 10^3.

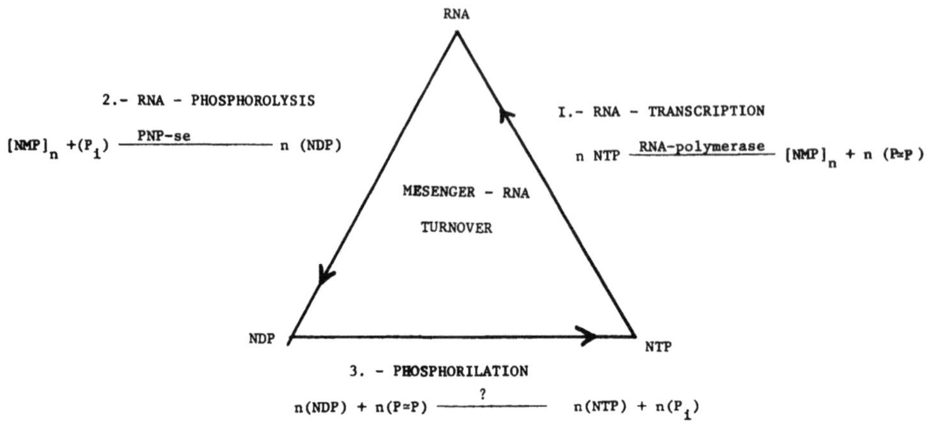

Figure 1. A "Closed" Reaction Cycle for m-RNA Turnover

Since the reaction of RNA transcription catalyzed by the DNA-dependent RNA-polymerase can be reversed by raising essentially the concentration of the pyrophosphates, these have to be hydrolized (or removed by another way from the system) to allow the reaction in the cell proceed in the direction of RNA synthesis.

The hydrolysis of pyrophosphates could provide the phosphate groups which are required to drive the reaction catalyzed by the PNP'se in the direction of phosphorolysis. It seems likely to suggest that the reaction of pyrophosphate hydrolysis is also coupled energetically with the 3rd step of the scheme, the NDP→NTP phosphorylation.

The general idea that pyrophosphate can replace ATP as an energy donor in some energy requiring biological reactions can be supported by solid experimental evidence. M. Baltscheffsky et al. (12) has shown that both PP_i and ATP may function as donors in such energy requiring reactions as those causing the changes in the redox state of endogenous cytochromes in chromatophores and in respiring mitochondria, both from lower and higher organisms. PP_i can act also as the energy donor in reactions responsible for the spectral shift of the carotenoid band in chromatophores (13). It was further shown (14) that the energy requiring reactions of transhydrogenase catalyzed reduction of $NADP^+$ by NADH in chromatophores may also be driven by the pyrophosphate instead of ATP.

However, all that evidence bears only on the coupling between the pyrophosphate hydrolysis and some oxido-reduction reactions. As far as we know, an enzyme which could catalyze the direct coupling of PP_i hydrolysis to the NDP→NTP phosphorylation has never been isolated and the reaction suggested as the 3rd step of the cycle in Fig. 1 has never been observed *in vitro*. Therefore, though the possibility of this way of utilizing the pyrophosphate macroergic bond energy is not excluded, we have to consider a more complicated possible way.

Kornberg *et al*. (15) have isolated and purified an enzyme from *E. coli* which catalyzes the reversible transfer of phosphate groups from polyphosphates to ADP:

$$(\text{Polyphosphate})_n + \text{ADP} \longrightarrow (\text{Polyphosphate})_{n-1} + \text{ATP}$$

The enzyme was obtained also from *Corynebacterium diphteriae*. It was demonstrated later that the formation of ATP from ADP and polyphosphates does really proceed in the cells of *Chlorobium thiosulfatorium* (16).

These data make plausible the assumption that *in vivo* the pyrophosphates liberated in RNA transcription (and perhaps in DNA replication too) are immediately condensed to polyphosphates and hence removed from the system. In the condensation reaction the orthophosphates, which are liberated, are those required for driving the phosphorolysis of m-RNA, and the energy is stored in macroergic bonds between the phosphate groups of the more stable polyphosphate molecule. The energy can be utilized later to close the reaction cycle of m-RNA turnover by the way of the reaction catalyzed by the Kornberg's enzyme.

Thus we came to the following hypothetical scheme of the reaction cycle performing the m-RNA turnover:

1. RNA transcription

$$n(\text{NTP}) \longrightarrow [\text{NMP}]_n + n(\text{P}{\sim}\text{P})$$

2. Pyrophosphate condensation

$$n(\text{P}{\sim}\text{P}) \text{-----------} (\text{Polyphosphate})_n + n\ \text{P}_i$$

3. RNA phosphorolysis

$$[\text{NMP}]_n + n\ \text{P}_i \longrightarrow n(\text{NDP})$$

4. NDP phosphorylation

$$n(NDP)+(Polyphosphate)_n \longrightarrow n(NTP)$$

The proposed reaction cycle is a closed one. It can rotate with a minimal "friction", i.e. with minimal energy losses. The information would be "cheap", as was predicted many years ago by F. Crick.

Acknowledgements

This study was carried as part of research program supported by a grant from the Jewish Agency, Jerusalem. We wish to thank Mr. Z. Voloch for helpful advices and Drs. Tatarskaya and Shoreq for the gifts of enzymes.

Summary

Heat effects of reactions of poly-A synthesis from ADP molecules and poly-A phosphorolysis, as well as for the reactions of ADP and poly-A hydrolysis, were measured and the values of ΔH for these reactions were computed. On the basis of data obtained a scheme was proposed of a closed reaction cycle performing the m-RNA turnover.

Additional Note

After the manuscript has been finished our attention was attracted to two papers in which the synthesis of ATP from ADP and pyrophosphate had been demonstrated to proceed in spinach chloroplasts (Bachofen, R., Lutz, H., Specht-Jürgensen Ingrid, FEBS Letters 1, 249 (1968) and in *Rhodospirillum rubrum* chromatophores (Keister, D., Minton, N.J., Biochem.Biophys.Res. Comm., 42, 932 (1971). That can be regarded as an evidence in favour of the mechanism proposed for the 3rd step of the cycle on Fig. 1 rather than of the reaction cycle involving the polyphosphates condensation. The question remains open, but it is possible that both ways are used by different cells and/or under different conditions.

References

1. Levinthal,C., Keynan,A., Higa,A.: Messenger RNA turnover and protein synthesis in B.subtilis inhibited by actinomycin D. Proc.N.A.S. 48, 1631-1638 (1962).

2. Levinthal,C., Fan,D.P., Higa,A., Zimmermann,R.A.: The decay and protection of messenger RNA in bacteria. Cold Spring Harbor Symposia on Quantitative Biology 28, 183-190 (1963).

3. Leive,L.: RNA degradation and the assembly of ribosomes in actinomycin-treated Escherichia coli. J.Mol.Biol. 13, 862-875 (1965).

4. Mangiarotti,G., Schlessinger,D.: Polyribosome metabolism in Escherichia coli. II. Formation and lifetime of messenger RNA molecules, ribosomal subunit couples and polyribosomes. J.Mol. Biol. 29, 395-418 (1967).

5. Kennell,D.: Titration of the gene sites on DNA by DNA-RNA hybridization. II. The Escherichia coli chromosome. J.Mol.Biol. 34, 85-103 (1968).

6. Salser,W., Janin,J., Levinthal,C.: Measurement of the unstable RNA in exponentially growing cultures of Bacillus subtilis and Escherichia coli. J.Mol.Biol. 31, 237-266 (1968).

7. Norris,T.E., Koch,A.L.: Effect of growth rate on the relative rates of synthesis of messenger, ribosomal and transfer RNA in Escherichia coli. J.Mol.Biol. 64, 633-649 (1972).

8. Georgiev,G.P.: On the structural organization of operon and the regulation of RNA synthesis in animal cells. J.Theor.Biol. 25, 473-490 (1969).

9. Tatarskaya,R.I., Lvova,T.N., Abrossimova-Amelyanchik,N.M., Korenyako,A.I., Bayev,A,A.: 5'-P-Forming exonuclease from Actinomyces sp. of coelicolor group, strain nr.5 (exonuclease A5). Purification and some properties. Eur.J.Biochem. 15, 442-449 (1970).

10. George,P., Rutman,R.J.: The "high energy phosphate bond" concept. Progress in Biophysics and Biophysical Chemistry 10, 2-53 (1960).

11. Peller,L.: In vitro RNA synthesis should be coupled to pyrophosphate hydrolysis. Biochem.Biophys.Res.Comm. 63, 912-916 (1975).

12. Baltscheffsky,M.: Inorganic pyrophosphate as an energy donor in photosynthetic and respiratory electron transport phosphorylation systems. Biochem.Biophys.Res.Comm. 28, 270-276 (1967).

13. Baltscheffsky,M.: Energy conversion-linked changes of caratenoid absorbance in Rhodospirillum rubrum chromatophores. Arch.Biochem. Biophys. 130, 646-652 (1969).

14. Keister,D.L.: - - - . Discussion following the paper: Baltscheffsky,M., Baltscheffsky,H., Stedingk,L.V.von: Light-induced energy conversion in the inorganic pyrophosphatase reaction in chromatophores from Rhodospirillum rubrum, in: Energy conversion by the photosynthetic apparatus. Brookhaven Symposia in Biology 19, 255-257 (1966).

15. Kornberg,A., Kornberg,S.R., Slimms,E.S.: Metaphosphate synthesis by an enzyme from Escherichia coli. Biochim.Biophys.Acta 20, 215-227 (1956).

Kornberg,S.R.: Adenosine triphosphate synthesis from polyphosphate by an enzyme from Escherichia coli. Biochim.Biophys.Acta 26, 294-300 (1957).

16. Cole,J.A., Hughes,D.E.: The metabolism of polyphosphates in Chlorobium thiosulfatophilum. J.gen.Microbiol. 38, 67-72 (1965).

3. Microbiological Calorimetry

3.1 Microcalorimetric Studies of Micro-Organisms

A.E. Beezer

It appears appropriate at this time to initiate a discussion on the directions in which microcalorimetric studies of microbiological systems are going. This will require an assessment of some published work and suggestions concerning some of the advantages and problems associated with this new methodology. During the development of the discussion some practical applications will be described as will some, as yet, unexplored possibilities.

Calorimetry is, as has often been observed, a general and non-specific reaction property. This property has been known and exploited for many years by chemists (1) but it is only since the introduction of true microcalorimeters (2) that the potential of microcalorimetric studies of microbiological systems has started to be explored. The generality of the method has led to the proposal (3) that microcalorimetry could be used to detect the presence of life on planets other than Earth. The proposal is based upon the fact that the complex sequence and conjunction of chemical events that is metabolism results in heat evolution and to the recording of a complex thermogram. The proposal was tested by incubating samples of Mexican desert dust (and specific organisms) in complex organic medium within a calorimeter.

These investigations forecast a more recent and general application of microcalorimetry namely the identification of micro-organisms (4, 5, 6). The reports describe over 200 clinically significant bacteria which it is claimed may be differentiated on the basis of their thermograms. The instrument used in these studies was a 50 channel batch type device employing sealed plastic ampoules. Typically thermograms show, initially, an apparent exponential increase in heat output rate followed by peaks and troughs. It is the magnitude and timing of these features which allows differentiation.

However the authors have attempted to broaden the utility of the technique by discussion of the feature detail of the thermograms in terms of specific metabolic events. Some reservations must be expressed about this extrapolation. It is possible that during the 7 - 20 hr timescale of the experiments the bacterial cells in a static, unstirred calorimeter may sediment. This may in turn induce many secondary effects. The oxygen tension (which is not controlled) will vary throughout the medium (incidentally obligate aerobes will not yield a thermogram under these conditions) as may all other properties of the system e. g. pH, substrate concentrations, osmotic tension etc. These purely physical events will/may have registrable heat effects. Overall therefore it would appear that thermogram detail will not easily be discussed in metabolic terms if the incubation has been conducted in sealed unstirred fermentors.

If though identification is the sole objective of the investigation then the limitations outlined above will not be important provided that the previous history of the cells does not influence the observed thermogram. There is preliminary evidence (7) that the source from which the organism is derived does indeed markedly influence the resultant thermogram. This point has, as yet, not been adequately explored.

Attempts to improve differentiation (perhaps only in the limited sense described above) have to date been based upon variation of carbon source e. g. changing glucose for lactose. It is possible, though, to control detail in the thermogram in another way namely by control of the oxygen tension (8) from zero to saturation. The magnitude and presence of peaks in the thermogram can be controlled, for example in Kluyveromyces fragilis, by varying the saturatin gas flow from N_2 to air to oxygen (Figs. 1, 2, 3).

This type of experiment has been performed in a flow microcalorimeter the operation of which has been described (9). To explain the detail in the

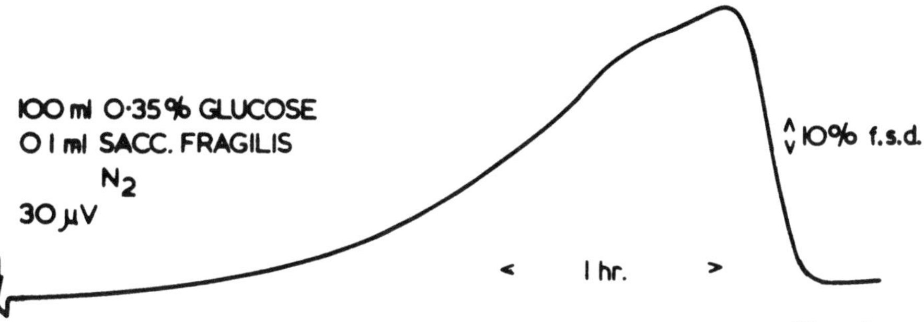

Fig. 1

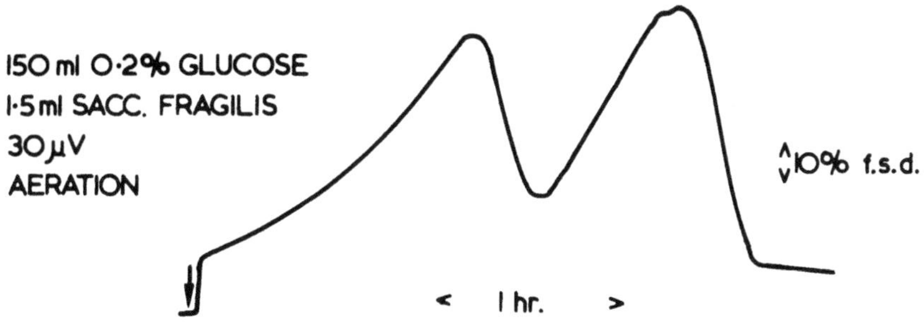

Fig. 2

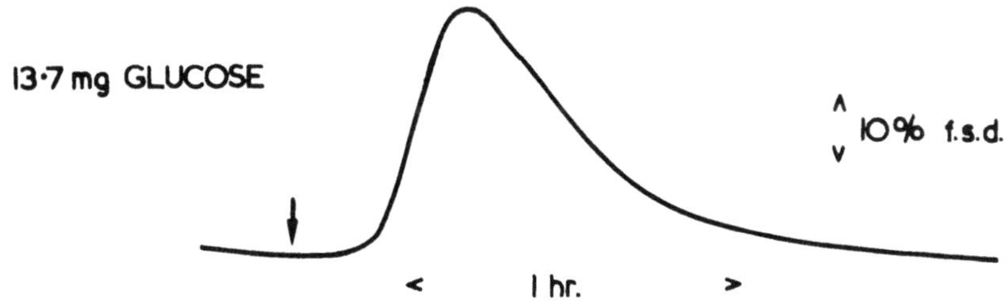

Fig. 3

thermogram requires the most careful chemical analysis to identify the disappearance of medium constituents and the appearance of metabolites as a function of time. Such experiments have yielded another cautionary observation - a thermogram containing a thermoneutral period which was accompanied by total depletion of all added oxygen. This means a reaction was proceeding but no evidence of its progress could be deduced from the thermogram. The reasons for this are not yet clear but the observation serves to illustrate that "no heat means no process" is not true.

On balance it would seem that absolute identification may not be possible but that characterisation may indeed be realisable. By this it is meant that in situations where strain maintenance is important e. g. bakery, brewery, cheese making etc. then calorimetry may provide a rapid, convenient and unique way of ensuring continuity of strain. This is because the background of the organisms may be regarded as constant. Moreover such simple, rapid experimentation may allow the investigation of alternative strains of potential industrial utility on the basis of thermogram type. Early evidence (10) suggests that yeasts suitable for differing commercial purposes show distinctive and process characteristic thermograms. This hypothesis remains to be thoroughly explored in detail. Again it is apparent that the thermogram detail arises, in part, from the decline in consumption of one component followed by the onset of metabolism of another medium constituent. This sequence of, say, sugar metabolism (see Fig. 4) is a traditional way (11) of speciating yeasts. Knowledge of the sequence and the rates of the processes involved are of importance in industrial processes.

Enlarging upon this theme of thermogram detail and medium constitution is should be obvious that the microcalorimetric approach allows a unique and hopefully sensitive approach to medium design. Many recipes for media involve up to 40 or so constituents of which many appear to be present for uncertain reasons. In addition variable components such as yeast extract, beef extract etc. appear in many medium designs.

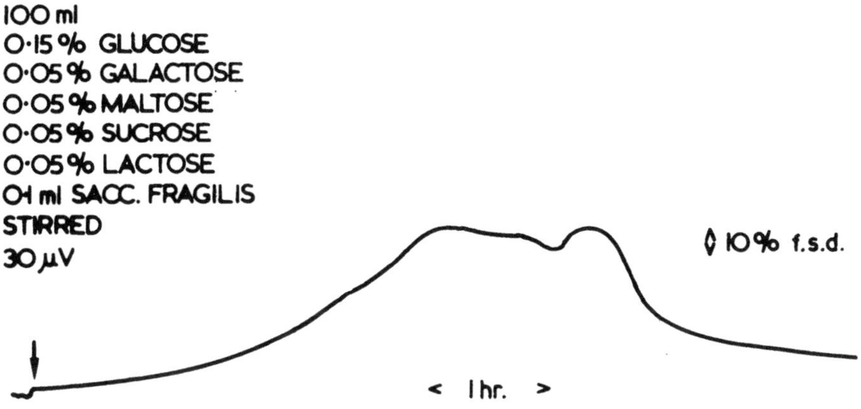

Fig. 4

It has been known for several years (12) that exponential heat output rate should accompany exponential growth in the incubation. Thus by observation of the effects on the thermogram of supplying or withdrawing metabolic nutrients one may investigate medium requirements - particularly important in this context are the trace components and vitamins. Such a series of experiments is illustrated in Figs. 5-7 where the effect of supplying limited vitamins (Figs. 6 and 7) can be compared with the thermogram ovserved when yeast extract is substituted for defined vitamin content (13).

Further possibilities exist in the uses of microcalorimetry in medium investigations such as the rapid investigation of the potential factors resulting in a commensal organism becoming pathogenic. Since by definition this involves a different metabolism then this effect should be observed in the thermogram.

Extending the discussion of metabolic modifiers further it should be apparent that antibiotic mode of action studies may usefully be researched. For example 5-fluorocytosine is an antifungal antibiotic which

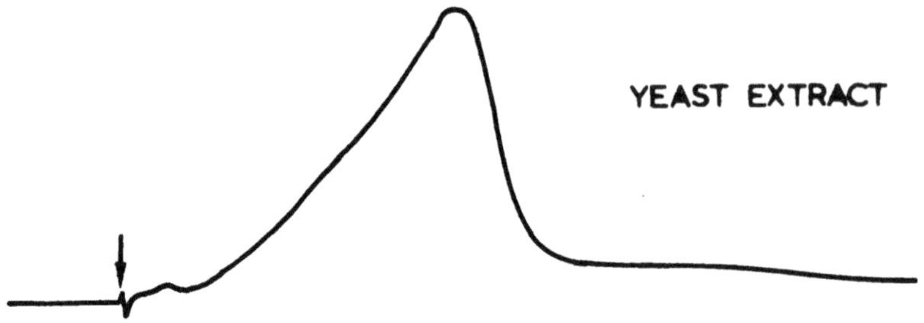

Fig. 5

Fig. 6

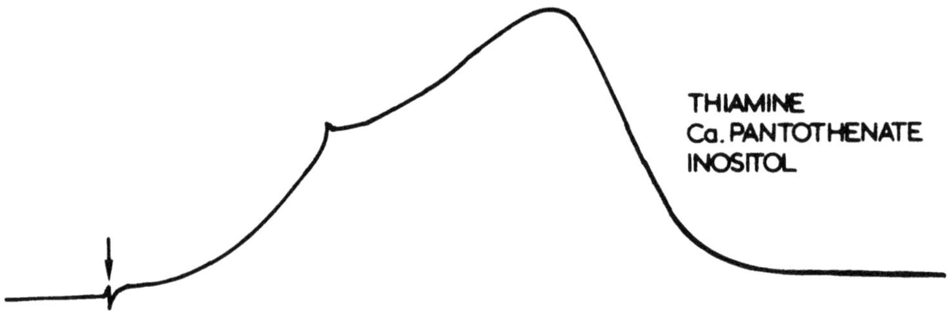

Fig. 7

has been shown to act (14) by interfering with RNA synthesis in contrast to the polyene fungals which act upon the membrane. This difference may be seen qualitatively (13) by comparison of the relevant thermograms (Figs. 8 and 9). The longer time required for action of the 5FC and its quasi-fungistatic appearance for several hours is strongly suggestive of a differing mode of action from the polyenes.

Fig. 8

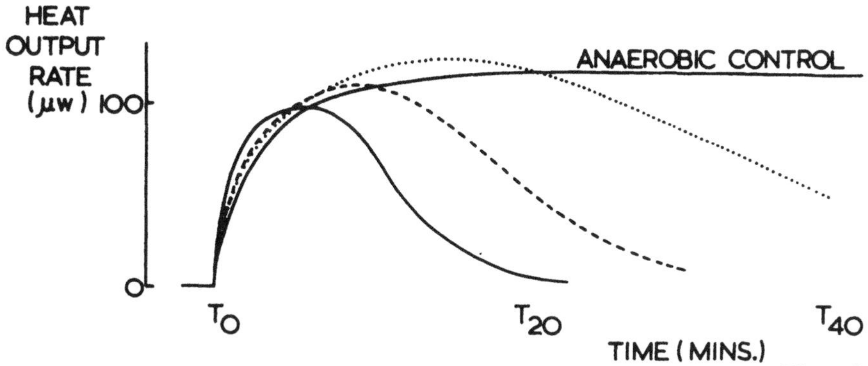

Fig. 9

From the observation of the effect of an antibiotic upon a growing culture of yeast it is but a short step to the investigation of a possible antibiotic assay based upon microcalorimetry. Such a technique has been established for the polyene antibiotics (15). The general features of the method are summarised in Fig. 9. Yeast cells stored in liquid nitrogen (9) when inoculated in to buffered glucose yield a zero order kinetic type reaction (illustrated as control in Fig. 9). In the presence of antibiotic the other curves are obtained. Measurement of the time for the thermogram to fall to some pre-determined level is the parameter recorded. Features of the technique compared to the classical agar plate diffusion technique are listed in the Table. The microcalorimetric assay is simple and rapid. It does not as yet have the throughput capacity of the agar plate diffusion method.

		Microcalorimetry	Agar plate diffusion
Reproducibility	%	3	5-10
Lowest determined concentration	unit/ml	0.1	20
Range	unit/ml	0.1-100	20-100
Time per assay	h	1	16

Inspection of Fig. 9 indicates that the thermogram contains information on the rate of the interaction of the antibiotic with the yeast cells. To date it has not been possible to deduce any simple relationship from these curves however the information is there. Likewise the figures illustrated earlier of growth processes in yeast contain kinetic data relating to the sequential consumption of medium constituents. Application of the known relationships (16, 17) to these situations has not been successful due to the complexity of the systems. However it must be one of the objectives of modern microbiological research to establish the formal relationships between the organism and the medium.

Conclusion

Inevitably a new methodology appears to find its initial application in the more qualitative and applied aspects of research. The potential of the microcalorimetric approach to the study of microbiological subjects in its quantitative aspects is as yet largely untouched. The possibility of deriving both thermodynamic and kinetic parameters from one experiment on such complex systems is appealing. The principal shortcoming at the present moment appears to be lack of precise analytical data to allow definition of the processes studied. Until precise thermodynamic equations can be written then the goals will prove elusive. Furthermore the interpretation of the kinetic detail will require a more complete understanding of adsorption, transport and reaction (as a minimum) processes.

References

1. Tyrrell, H. J. V.,Beezer, A. E: Thermometric Titrimetry, Chapman and Hall, London (1968)
2. Monk, P. Wadsö, I.: A flow micro reaction calorimeter, Acta Chem. Scand. 22, 1842-1852 (1968)
3. Bechman Inst. Inc., Contract No. NAS 2-3477, Final Report (1966)
4. Boling, E. A., Blanchard, G. C., Russell, W. J.: Bacterial identification by microcalorimetry, Nature 241, 272-273 (1973)
5. Russell, W. J., Farling, S. R., Blanchard, G. C., Boling, E. A.: Interim review of microbial identification by microcalorimetry, in: Microbiology - 1975 Ed.: Schlessinger, D., Am. S. Microbiol., Washington (1975)
6. Russell, W. J., Zettler, J. F., Blanchard, G. C., Boling, E. A.: Bacterial identification by microcalorimetry, in: New Approaches to the Identification of Microorganisms, Ed. Hegen, C.-G., Illeni, T., Witey, New York (1975)
7. Beezer, A. E., Bettelheim, K. A., Shaw, E. J., unpublished observations (1976)
8. Newell, R. D.: Bioassay of Nystatin, Ph. D. Hesis, London University (1975)
9. Beezer, A. E., Newell, R. D., Tyrell, H. J. V.: Application of flow microcalorimetry to analytical problems: The preparation, storage and assay of frozen inocula of Saccharomyces cerevisiae, J. Appl. Bacteriol. 41, 197-207 (1976)
10. Beezer, A. E., Newell, R. D., Tyrrell, H. J. V.: manuscript in preparation (1976)
11. The Yeasts, Ed. Lodder, J., North Holland, Amsterdam (1971)
12. Forrest, W. W.: Bacterial microcalorimetry, in: Biochemical Microcalorimetry, Ed. Brown, H. D., Academic Press, London (1969)

13. Beezer, A. E., Newell, R. D., Cawson, R. A.: unpublished observations (1976)

14. Beggs, W. H., Sarosi, G. A., Steell, N. M.: Inhibition of potentially pathogenic yeastlike fungi by clotrimazote in combination with 5-fluorocytosine or Amphotericin-B, Antimicrobial Agents and Chemotherapy. 2, 863-871 (1976)

15. Beezer, A. E., Newell, R. D., Tyrell, H. J. V.: Application of flow microcalorimetry to analytical problems. The bioassay of nystatin bulk material, Analyt. Chem., in press (1976)

16. Beezer, A. E., Tyrrell, H. J. V.: Application of flow microcalorimetry to biological problems. Part 1. Theoretical Aspects, Science Tools, 19, 13-16 (1973)

17. Biltonen, R. L., Johnson, R. E.: Determination of reaction rate parameters by flow microcalorimetry, J. Am. Chem. Soc. 25, 2349-2355 (1975)

Dr. A. E. Beezer
Chemistry Department
Chelsea College
University of London
Manresa Road
London SW 3 6 LX, U. K.

3.2 Calorimetric Studies of Yeast Metabolism under Nongrowing Conditions

T. Fujita, K. Nunomura

INTRODUCTION

Calorimetric studies on growth and metabolism in yeast have been systematically carried out by several workers (1,2,3), however, many of the details of the thermogram still remain unsolved in relation to the individual processes of metabolism. For the purpose of this study, it is often more advantageous to use nongrowing conditions, because metabolic processes then occur without involving complicating factors of growth. The experimental results can be analyzed after several hours. This paper describes a calorimetric study on some metabolic processes in yeast under nongrowing conditions by using specially designed calorimeter for microbiological processes.

MATERIALS AND METHODS

Calorimeter and experimental procedure

The calorimeter and experimental procedure used have been described in details elsewhere (4). The apparatus is a rotation calorimeter of the conduction type, with a maximal sensitivity of 30 mm deflection of recorder pen for a heat flow of 1 cal/hr and the time constant of the response is 8 min. The culture vessel is a short-necked cylindrical flask made of Pyrex glass with a capacity of 40 ml and is usually charged with 5 ml of sample in order to assure efficient aeration and agitation.

---

†Present address: Thermochemistry Laboratory, Chemical Center, Lund University, S-220 07 Lund, Sweden

The inoculum previously separated by a partition wall was mixed with the medium by starting the rotation and the reaction in the active vessel was compared with a control system in the reference vessel. During measurements the calorimeter was rotated continuously.

The culture was aerated by using an aeration assembly in which moist air is pumped in at a rate of 15 ml/min and after passing through a heat exchange unit, the air fills the space between the lide and heat sink of the calorimeter and spontaneously enters the culture vessel via cotton-wool plugs. Excess air and exhaust gas are allowed to flow through outlet tubes. The anaerobic experiments were performed under an atmosphere of nitrogen.

Organism and culture

The yeast studied (Saccaromyces cerevisiae) was a parent strain of a respiratory-deficient strain kindly provided by Dr. T. Gunge.† The strain was maintained on agar slants and grown in Kilkenny and Hinshelwoods synthetic medium (5). After 16 hours cells were harvested by centrifugation, washed three times with distilled water and suspended in potassium phosphate buffer (pH 5.6). Unless otherwise noted, calorimetric experiments with the washed suspension were carried out under aerobic conditions.

Chemical analyses

Since it is difficult to remove samples from the culture vessel without any thermal disturbance to the calorimetric measurement, a parallel culture for observation and analysis was used. The culture was contained in a calorimeter vessel and maintained in a similar way.

---

†Mitsubishi Chemical Industries Research Laboratory

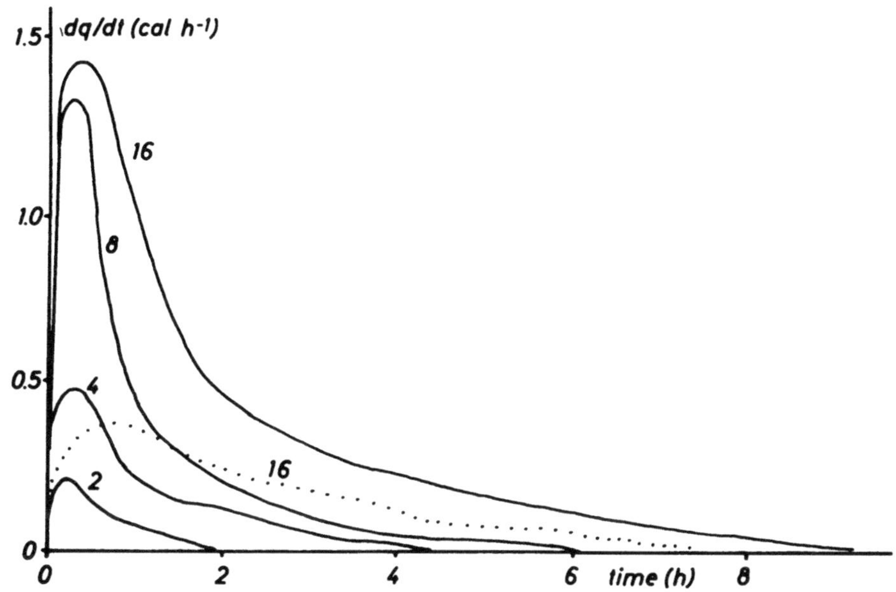

Fig. 1. Time course of endogeneous heat production by washed cells of S. cerevisiae with various amounts of cells. ——, under aerobic conditions, ····, under anaerobic condition. Figures indicate the dry weight of the cells. Cells suspended in phosphate buffer (0.5 ml) were diluted with phosphate buffer (4.5 ml).

Glucose was determined enzymatically using the Glucostat method and ethanol was determined by gas chromatography. Turbidity of the culture was measured at 550 nm with a Shimazu Spectronic 20 spectrophotometer by using 0.1 ml microcuvettes.

RESULTS AND DISCUSSION

Endogeneous heat production

A considerable amount of heat is produced when a suspension of S.cerevisiae is added to a suspending medium in the absence of exogeneous energy source. Fig. 1 shows the thermograms obtained when different suspensions of yeast were added to a phosphate buffer. The heat was produced under anaerobic conditions, showing the existence of anaerobic processes in this microbe. The total amount of heat calculated from the thermogram was

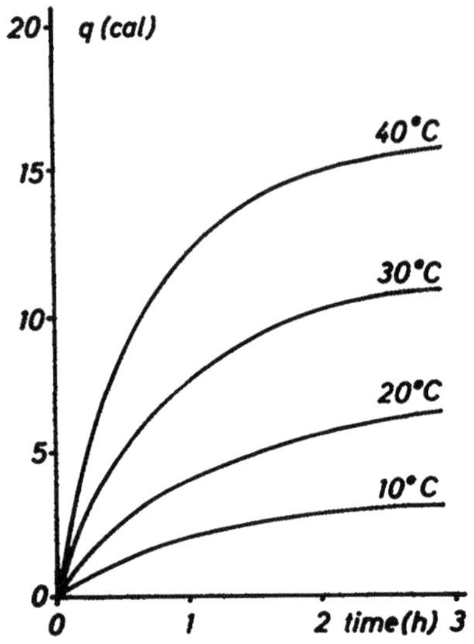

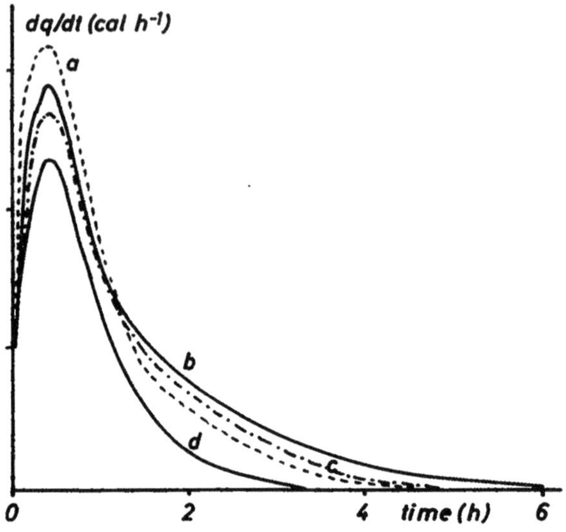

Fig. 2. Effects of temperature on the endogeneous heat production.

Fig. 3. Effect of suspending media on the endogeneous heat production. a, cells suspended in 0.2 M/L NaCl solution, diluted with distilled water; b, cells suspended in distilled water (pH 4.7), diluted with water adjusted to pH 7 by NaOH, resulting pH was 6.1; c, cells suspended in distilled water (pH 4.7), diluted with water adjusted pH 3.1 by HCl, resulting pH was 3.3; d, cells suspended in buffer, diluted with buffer. In these experiments the reference vessel contained a similar reaction, but without cells, eliminating the effect of the suspending media.

proportional to the cell concentration. This relationship, considered together with the fact that starvation markedly affected the heat production, indicates that this heat production depends on the reserve of intracellular energy sources. Although Forrest (6) reported that lower concentrations of cells of Streptococcus faecalis produce heat at a greater rate per unit dry weight than the higher concentrations there was no difference in rate between samples of different cell density in the present experiments.

Fig. 2 shows the time course of the heat production at various temperatures. There is a linear relationship between the total amount of heat and temperature up to 40°C. This simple temperature characteristic supports the view that in this temperature range, heat production may be mainly due to the same processes. It is shown in Fig. 3 that the ionic property of suspending media affects the thermogram. There was no difference in oxygen uptake between these samples, however, the surface of yeast cells have some iogenic groups that have ion-exchange properties, the difference in the heat production therefore may be due to the physico-chemical reaction of yeast cells with suspending media rather than to their physiological response.

The results obtained in the present experiments suggest that this heat production may be due to the endogeneous metabolism of the cells adjusting their rate of metabolism in accordance with the change in their environmental condition. It seems to be analogus to the heat of dilution first reported by Forrest (7) and later by Ross (8). As washed cells of <u>S. cereviasiae</u> have a higher level of endogeneous metabolism than that of bacteria, special attention should be paid to this heat production when interpreting the initial part of the thermogram.

Heat production of glycolysis

When an exogeneous substrate is added to a washed suspension of yeast cells the heat production subsequently observed may have components due to glycolysis and endogeneous metabolism. Fig. 4 shows the time course of aerobic glycolysis by yeast in which turbidity, glucose and ethanol concentrations found in the culture apparatus are compared with a corresponding thermogram obtained with the calorimeter. Glucose was taken up within one hour and ethanol formed by the glucose fermentation was exhausted after about 6 hours. The small increase in turbidity

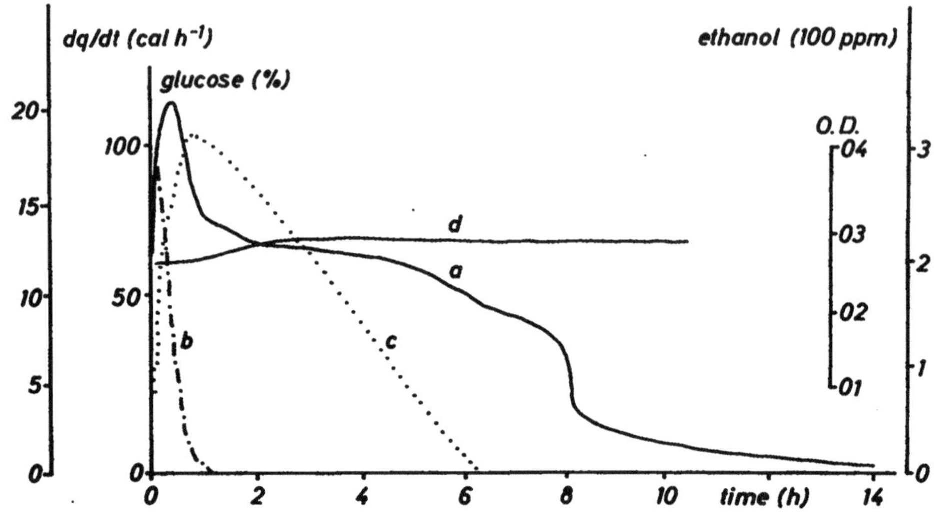

Fig. 4. Time course of the aerobic glycolysis by washed cells of S. cerevisiae. a, thermogram; b, glucose concentration; c, ethanol concentration; d, turbidity. Cells (8 mg dry weight) suspended in buffer were added to the buffer containing glucose (0.2 % V/V).

was not due to the cell multiplication but to the increase in cell volume as examined by microscopic observation. The thermogram shows three distinct phases. From the results of glucose analyses the first phase seems to be associated with glucose uptake. The results of the ethanol analyses suggest that the second phase is due to the consumption of ethanol, which is followed by the third phase of endogeneous metabolism.

To elucidate the nature of the first phase observed in the thermogram of glycolysis, the following experiments were carried out in which either glucose was replaced by non-metabolizable sugars or glycolysis was inhibited by inhibitors. S. cerevisiae cannot ferment 3-O-methyl glucose, galactose and sorbose, however, substantial heat was produced by the addition of these sugars (Fig. 5). The thermograms are different from that with glucose, showing a single broad peak with a smaller heat production.

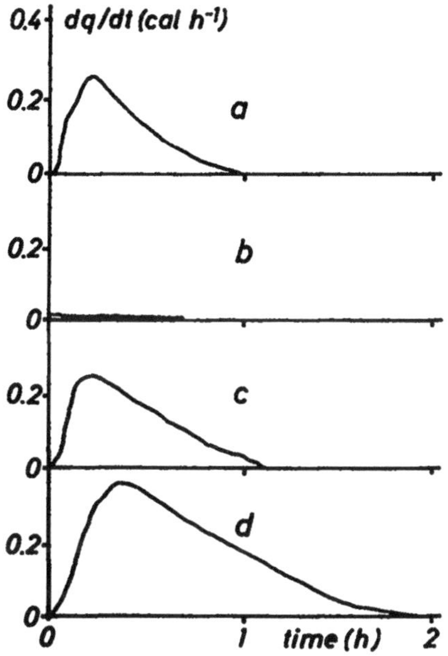

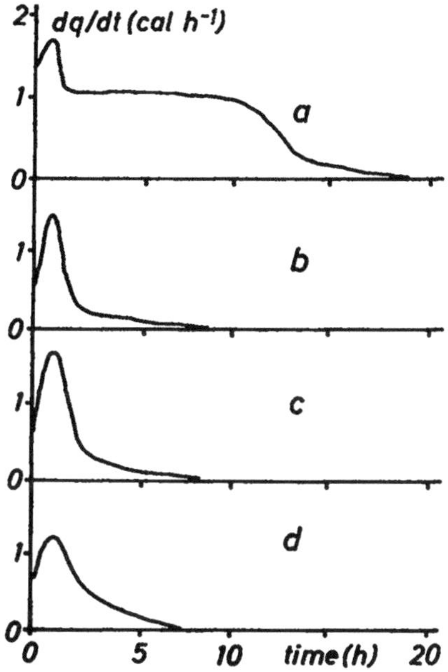

Fig. 5. Thermograms for the uptake of non-metabolizable sugars. a, 3-O-methyl glucose under aerobic condition; b, 3-O-methyl glucose under anaerobic condition; c, galactose; d, sorbose.

Fig. 6. Effects of metabolic inhibitors on the thermogram of glycolysis. a, control; b, with 750 μmole KCN; c, under anaerobic condition; d, with 60 μmole CCCP.

Chemical analyses show that considerable amounts of these sugars were taken up by yeast. As shown in Fig. 6 metabolic inhibitors have marked effects on the thermogram. The first phase appears to take a similar pattern in each of the curves, but the subsequent heat production is hardly observed. From these results it seems probable that the first phase of the thermogram of aerobic glycolysis is mainly associated with the uptake of glucose. The time course of the glycolysis suggests that at an early stage, glucose is taken up via a mechanism coupled to the endogeneous metabolism and the second phase depending on the glycolysis may begin at nearly the same time as glucose uptake. The thermograms shown in the present paper have not been reported by other investigators, probably because the two phases overlap and cannot be distinguished separately at a relatively low cell concentration and a high glucose concentration.

SUMMARY

A considerable amount of heat was produced when a suspension of yeast was added to a suspending medium in the absence of an exogeneous energy source. The amount of heat depended on cell concentration, temperature, and environment of the microbe. Characteristic patterns of heat production were found when suspensions of yeast were added to phosphate buffer containing glucose. Experiments with non-metabolizable sugars and the effects of inhibitors suggested that the early phase of the reaction was mainly associated with the uptake of glucose.

REFERENCES

1. Battley, E.H.: Enthalpy change accompanying the growth of Sacchromyces cerevisiae. Physiol. Plant. 13, 628-640 (1960).
2. Lamprecht, I., Schaarschmidt, B., Stein, W.: Mikrokalorimetrische Untersuchungen zum Stoffwechsel von Hefen. III. Betriebsstoff wechsel in Glucosepuffer. Biophysik 10, 177-186 (1973).
3. Murgier, M., Belaich, J.P.: Microcalorimetric determination of the affinity of Saccharomyces cerevisiae for some carbohydrate growth substrates. J. Bacteriol. 105, 573-579 (1971).

4. Fujita, T., Nunomura, K., Kagami, I., Nishikawa, Y.: Design and testing of a calorimeter for microbiological uses. J. Gen. Appl. Microbiol., 22, 43-50 (1976).

5. Kilkenny, B.C., Hinshelwood, S.C.: Proc. Roy. Soc. B London, 138, 375 (1951).

6. Forrest, W.W.: Bacterial Calorimetry. In: Biochemical microbiology, ed. by Brown, H.D., Academic Press, New York, 165-180 (1969).

7. Forrest, W.W., Walker, D.J.: Calorimetric measurements of energy of maintenance. Biochem. Biophys. Res. Commun. 13, 217-222 (1963).

8. Ross, P.D., Fletcher, A.P., Jamieson, G.A.: Microcalorimetric study of isolated blood platelets in the presence of thrombin and other aggregating agents. Biochim. Biophys. Acta 313, 106-118 (1973).

3.3 Microcalorimetric Measurements of the Heat Production in Partially Synchronous Cultures of Baker's Yeast

R. Brettel

The energy metabolism of microorganisms has often been studied with calorimetric methods. All these studies, however, have the difficulty of relating the measured heat effect to a corresponding event on the level of the cells or the culture. This is due to the non-specifity of calorimetry. Usually we need a variety of additional physical, chemical and biological parameters to make calorimetric results more significant, i.e. to relate them to determined external culture conditions or to average physiological cell conditions. But most of these investigations neglect the individual state of the cells, since growing cultures contain a heterogenous mixture of cells in every age of the cell cycle.

The aim of our investigations is to relate the turnover of heat to the events of the cell division cycle. For these experiments we used the budding yeast Saccharomyces cerevisiae, since the appearance and growing of the bud is a convenient visual marker of the position of the cell in the dividing cycle.

The sequence of the most important events in the cell cycle of baker's yeast has been well studied [1,2]. The cycle in budding yeast (Fig.1) starts with a single unbudded cell in the G_1 interval.

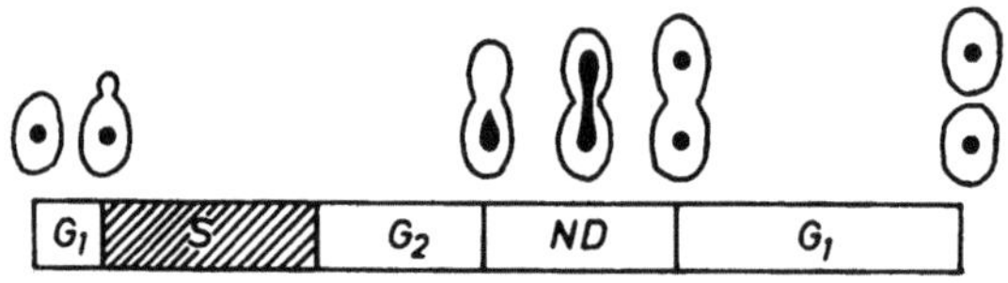

Fig.1: Cell cycle of Saccharomyces cerevisiae (adopted from [2])

The end of this interval is marked by the initiation of DNA synthesis (S) and the emergence of the bud, which grows in size throughout the cycle. Nuclear division (ND) and the migration of a nucleus into the bud take place in the second half of the cycle. After cell wall separation the cycle is completed with the production of two unbudded cells.

Preparation of Synchronous Cultures

Most of the methods for analyzing the cell cycle are based on examination of either single cells or synchronous cultures. Calorimetric work with single cells is nearly impossible, owing to the small heat production of the cell and the too low sensitivity of the instruments. Therefore synchronously dividing cultures are necessary. There are two main ways of making synchronous cultures: selection and induction [2]. In selection methods uniform cells are separated physically from a heterogenous culture and grown as a synchronous culture. In induction methods a whole culture is induced to divide synchronously by physical or chemical treatments.

We induced partially synchronous growth in a continuous chemostat culture with a method described in [3]. In the chemostat the cell culture is fed with fresh medium at a constant rate and the same amount of culture suspension flows out of the vessel. This way the culture comes to a steady-state during which all important parameters of the culture remain constant. Nevertheless, continuous cultures may show metabolic oscillations which are due to partially synchronous growth. The mechanism of this synchronisation is not well understood at present but it can be induced by starvation or occurs spontaniously [4]. Fig.2 shows the stable oscillations of heat evolution in a continuous culture which are produced by changes in the metabolic activity due to synchronous growth. The method produces 30 to 50% synchrony at a rather high cell concentration. However, it should be considered that the cells grow under special physiological conditions, i.e. under a lack of exogenous substrate.

During the appearance of synchrony oscillations we determined by microscopic inspection the percentage of initial budding cells over several cell cycles. Although the definition of initial budding cells may cause some subjective errors, we found this method most successful to determine the cycle position of synchronously dividing cells.

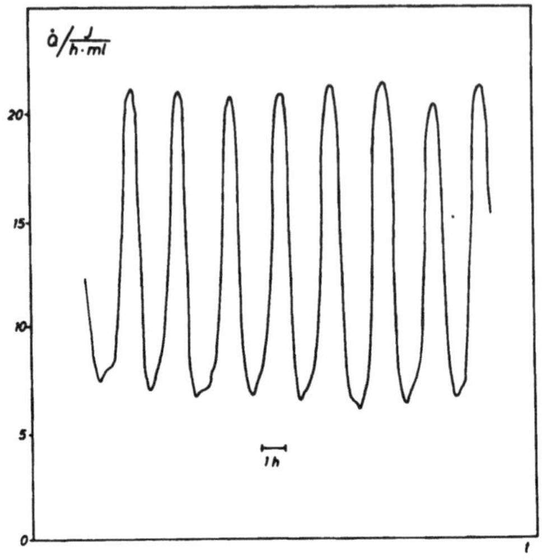

Fig.2: Thermogramme of a partially synchronous chemostat culture of Saccharomyces cerevisiae

Fermentor System and Calorimeter

In our experiments we used a combination of a fermentor system and a flow-microcalorimeter [5,6]. Fig.3 shows a simplified scheme of this device. The culture is operated outside the calorimeter and only small parts of the culture volume are pumped continuously through the calorimeter. With this method it is easily possible to control the cultivation conditions and to measure additional parameters during the run of the culture. For the measurements we used a heat conduction microcalorimeter (LKB) with a tubular flow cell. The fermentor system is an aerated and stirred 1 liter laboratory fermentor (BIOTEC) equipped with electrodes and devices to control pH-values and to measure oxygen concentration in the culture.

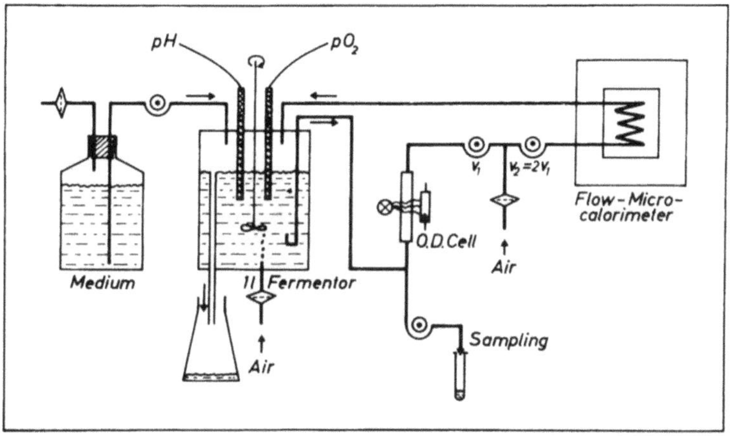

Fig.3: Scheme of the arrangement of a chemostat combined with a flow-microcalorimeter

The cell suspension from the fermentor is pumped continuously through a device to measure the turbidity and thereafter through the calorimeter. To avoid sedimentation of the yeast cells and to guarantee sufficient oxygen supply in the calorimeter vessel and in the tubing, the flow stream is mixed half and half with air. This arrangement, however, produces a noise which interferes with the real signal of the calorimeter. Therefore we had to damp the voltage output with a capacitor which limits the device to slow processes. Another peristaltic pump is used to take samples for the determination of dry mass, substrate or metabolite concentration and oxygen consumption with a polarographic method. The equipment for the continuous culture is a conventional chemostat. Fresh medium is supplied to the fermentor vessel at constant rates with a peristaltic pump, and the same volume of the cell suspension leaves the system through an overflow tube.

Results and Discussion

Fig.4 gives the results of a synchronously growing culture with glucose as limiting substrate. It shows the part of a thermogramme

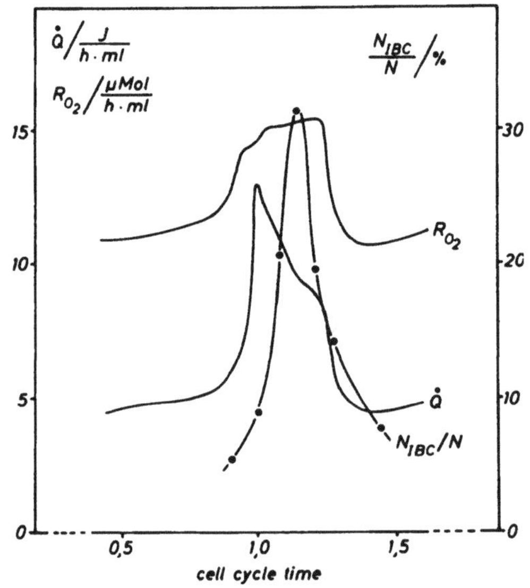

Fig.4: Glucose limited chemostat culture of Saccharomyces cerevisiae. Percentage of initial budding cells N_{IBC}/N, rates of heat generation $\dot{Q}$ and of oxygen consumption R_{O_2} as functions of cell cycle time. Cell cycle time is given in fractions of a complete cell cycle. For convenience the cycle's start and stop were set at the maximum values of heat evolution.

which represents a complete cell cycle. The time is indicated in fractions of a complete cell cycle. At a sudden point the rate of heat evolution $\dot{Q}$ increases steeply to a 2.5 fold value. The rate of oxygen consumption R_{O_2} likewise rises, but without the characteristic peak of the thermogramme. Parallel to these values the percentage of initial budding cells N_{IBC}/N shows a strongly marked maximum of about 30%. This means that the activity of the metabolism increases considerably a short time before the bud appears. Accordingly the cell has to achieve a high rate of biosynthesis and of structural changes to prepare for the doubling process and the emergence of a bud within a short time. The required energy is released by the degradation of reserve carbohydrates, which are stored as trehalose and glycogen inside the cell [3,4].

It is proved that the endogenous carbohydrate reserves are mainly catabolized via the same pathways as exogenous substrates [7]. The increase in oxygen consumption indicates that a large amount of the intensified metabolism is due to oxidative degradation. During the time of most intense metabolism, however, the degradation increases so much that the capacity of the Krebs cycle is too small. Part of the reserves is then catabolized only fermentatively to ethanol, which is secreted into the medium [3]. This becomes evident when we look on the quotient of heat generation and oxygen consumption (Fig.5). For the oxidative degradation of carbohydrates we have a regular value of about 0.47 J/μMol O_2 [8]. As we find higher values here, this means that part of the evolved heat is not released by oxidative processes.

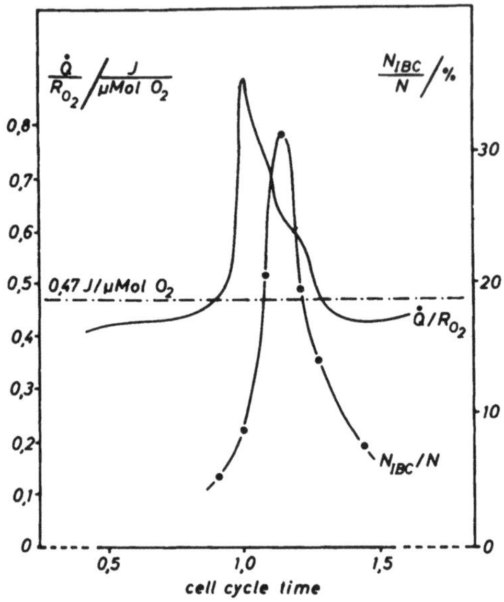

Fig.5: Ratio of heat generation and oxygen consumption $\dot{Q}/R_{O_2}$ as function of cell cycle time, calculated from data in Fig.4. The level at 0,47 J/µMol O_2 represents an average value for the oxidative degradation of carbohydrates.

We obtained similar results from synchronous yeast cultures which grow on ethanol as sole and limiting substrate (Fig.6). The degree of synchrony is lower but the relation of initial budding cells and heat production shows essentially the same characteristics. The rate of oxygen consumption remains nearly constant. This suggests that again part of the required energy is released from fermentative degradation of reserve carbohydrates, which were synthesized before via gluconeogenesis. Although there is no fermentable exogenous substrate in ethanol limited cultures, fermentation of endogenous reserves takes place.

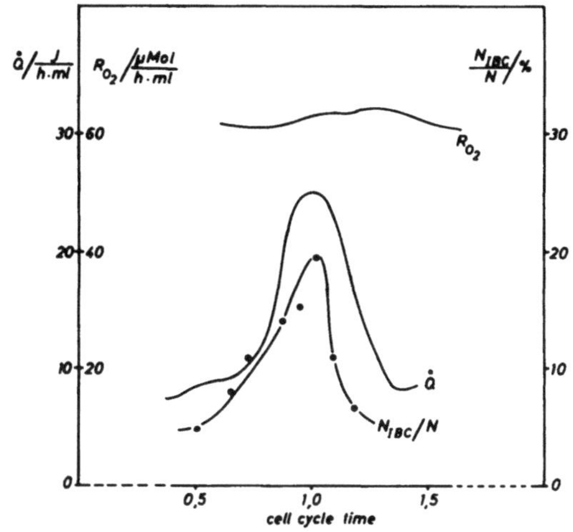

Fig.6: Ethanol limited chemostat culture of Saccharomyces cerevisiae. Plotting as in Fig.4.

As a further result of these experiments, we found that the period of increased metabolism has a nearly constant length, which is independent of the specific growth rate and the duration of the complete cell cycle. This observation corresponds with results reported in [9] according to which the time from the emergence of the bud till to the separation of the complete cell remains constant. A prolongation of the cell cycle therefore implies a prolongation of the single cell stage only and not of the cell division process itself.

References

1. Hartwell, L.H.: Saccharomyces cerevisiae cell cycle. Bacteriol. Rev. 38, 164-198 (1974).

2. Mitchison, J.M., Carter, B.L.A.: Cell Cycle Analysis. In "Methods in Cell Biology" (D.M. Prescott, ed.), Vol. XI, pp. 201-219. Academic Press, New York 1975.

3. Küenzi, M.T., Fiechter, A.: Changes in Carbohydrate Composition and Trehalase-Activity During the Budding Cycle of Saccharomyces cerevisiae. Arch. Mikrobiol. 64, 396-407 (1969).

4. Küenzi, M.: Über den Reservekohlenhydratstoffwechsel von Saccharomyces cerevisiae. Dissertation, ETH, Zürich, 1970.

5. Brettel, R.: Microcalorimetric measurements of the energy utilization in chemostat cultures of Saccharomyces cerevisiae. In: "Proc. Fourth Intern. Symp. on Yeasts" (H. Klaushofer, U.B. Sleytr, eds.), Vol. 1, pp. 87-88. University of Agriculture, Wien 1974.

6. Brettel, R., Corti, L., Lamprecht, I., Schaarschmidt, B.: Combination of a continuous culture with a flow-microcalorimeter. Stud. Biophys. 34, 71-76 (1972).

7. Eaton, R.E.: Endogenous metabolism of reserve deposits in yeast. Ann. N.A. Acad. Sci. 102, 678-687 (1963).

8. Minkevich, I.G., Eroshin, V.K.: Productivity and Heat Generation of Fermentation Under Oxygen Limitation. Folia microbiol. 18, 376-385 (1973).

9. Meyenburg, H.K.v.: Der Sprossungszyklus von Saccharomyces cerevisiae. Path. Microbiol. 31, 117-127 (1968).

3.4 Quantitative Relation between Heat Production and Weight during Growth of Microbial Cultures

B. Schaarschmidt, A.I. Zotin, I. Lamprecht

Summary: The proportionality between heat production and the logarithm of the body weight during growth, found valid for higher organisms [1], is checked experimentally for microorganisms (Saccharomyces cerevisiae, Bacillus stearothermophilus). On comparing data in the literature we found different constants of proportionality (Transition coefficient r) for different organisms under different growth conditions. The results are discussed in view of different efficiencies of metabolism.

Introduction

The theory of Irreversible Thermodynamics [2] assumes direct proportionality between the dissipation function ψ (specific rate of entropy production) and the specific rate of heat production $\dot{q} = \frac{dq}{dt}$:

$$\dot{q} \sim \psi = \sum_j J_j \cdot X_j \qquad (1)$$

where J_j = thermodynamic flow and X_j = thermodynamic force. One of the basic concepts of this theory requires a linear dependance between the flows and the forces

$$J_i = \sum_j L_{ij} \cdot X_j \qquad (2)$$

(L_{ij} = phenomenological coefficients).

If the theory of irreversible thermodynamics is applied to living matter, eq.(2) must hold for the growth of an organism too. In a first approximation growth may be assumed to be determined by one force only. Then the growth equation can be written as

$$\bar{J}_g = L_{gg} \cdot X_g \qquad (3)$$

($\bar{J}_g = \frac{Jg}{P}$ = specific flow of weight change, index g = growth).

Assuming a phenomenological law of growth [1]

$$\frac{1}{P} \cdot \frac{dP}{dt} = k \cdot (T_m - t) \qquad (4)$$

(P = dry weight, k = constant, T_m = time when P has reached maximum) which has been found valid for growth of animals and man, we obtain

$$\bar{J}_g = \frac{1}{P} \cdot \frac{dP}{dt} = k \cdot (T_m - t)$$

$$\text{and} \quad X_g = (T_m - t) \;. \qquad (5)$$

Under consideration of eqs.(5) and (6) the dissipation function (1) can be written as

$$\dot{q}_g = r \cdot k \cdot (T_m - t)^2$$

which can be finally completed if $(T_m - t)^2$ is inserted from the solution of the law of growth (4)

$$\dot{q}_g = r \cdot 2 \cdot \log e \cdot \log \frac{P_m}{P} \qquad (6)$$

(r = transition coefficient given in mW/g).

In a calorimetric experiment on the heat production of a growing organism only the total rate of heat production is recorded, which is generated by the basic metabolism $\dot{q}_o$ as well as by the growth metabolism $\dot{q}_g$. With $\dot{q}_g$ from eq. (6) follows

$$\dot{q} = \dot{q}_o + \dot{q}_g$$

$$= \dot{q}_o + r \cdot 2 \cdot \log e \cdot \log \frac{P_m}{P} , \qquad (7)$$

and putting $A = \dot{q}_o + 2r \cdot \log e \cdot \log P_m$; $B = 2r \cdot \log e$, yields

$$\dot{q} = A - B \cdot \log P . \qquad (8)$$

By equation (8) a semilogarithmic dependance between the specific rate of heat production and the body weight in a growing organism is stated, contrary to the generally assumed logarithmic dependance (allometric function)

$$\log \dot{q} = a - b \cdot \log P \qquad (9)$$

as was confirmed between the intensity of respiration and weight [1]. The validity of (8) was verified by calorimetric experiments for growth of fish, frog and fowl [3], but not with unicellular organisms. Since to a certain extent there are some similarities between a single multicellular organism and a culture composed of many separate cells, it was considered desirable to prove eq. (8) with respect to microorganisms.

Material and Methods

Organisms: In most experiments the diploid yeast strain 211 of *Saccharomyces cerevisiae* was used, but in some other experiments related polyplonts (haploid to hexaploid [4]) and the thermophilic bacterium *Bacillus stearothermophilus* were also used. The growth medium contained yeast extract (10 g/l), peptone (5 g/l) and glucose (5 to 20 g/l).

Calorimeter: The rate of heat production was measured in two different Calvet-microcalorimeters (Setaram/France) with a sensitivity of 50 µV/mW and 100 ml working volume or 60 µV/mW and 10 ml, respectively. In the calorimeter vessel the cultures were stirred under air. The temperature was kept at 30°C for yeast and at 65°C for *B.stearothermophilus*. The heat flux during growth is recorded continuously as a thermogram. A detailed description of the culturing conditions and the calorimetric method was given previously [5].

Weight determination: Two methods were used for calculating the dry weight. With one method each growth experiment in the calorimeter was stopped at another time, the grown mass sampled and determined by drying and weighing. In the second method the optical density OD of the culture was measured by light guides in the calorimeter vessel simultaneously with heat flux [6] and converted to a scale of dry weight using the direct proportionality: 1 mg/ml ≙ 0.32 OD units for diploid yeast and 0.59 units for *B.stearothermophilus*.

Mathematical calculation: Values for the rate of heat production and for the dry weight were taken from the experimental diagrams with a curve analyzer (Stange u. Voss/Germany). By a regression analysis after Pearson the slope B, the transition factor r and the correlation factor for equation (8) were determined.

Results

Fig.1 gives typical thermograms for growing cultures of the organisms used. After an initial lag-phase due to adaption of the organism to the medium, the heat production rises according to the

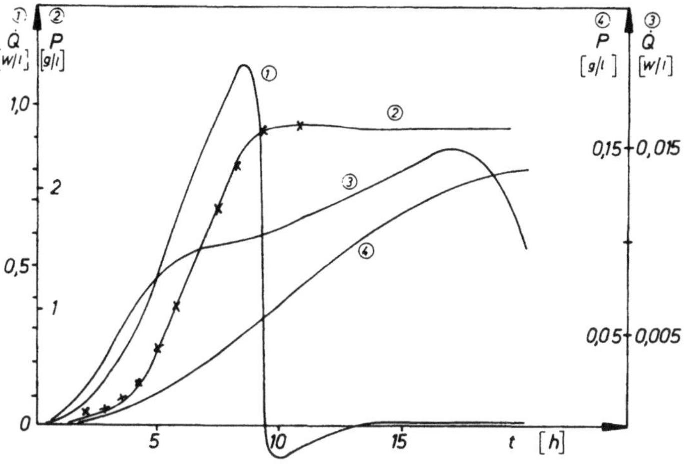

Fig.1: ① Thermogram of a growing culture of Saccharomyces cerevisiae at 30°C, ② corresponding biomass P. ③ Thermogram of Proteus mirabilis (after [9]), ④ corresponding biomass. x = calculated from eq.(4).

exponential growth of the culture. The heat flux reaches its maximum shortly before the energy source (glucose) is consumed and then drops rapidly. For the calculation of equations (8) and (9) only the exponential growth phase was analyzed as nearly all growth took place during this phase. Figure 2 gives typical results in semilogarithmic and Figure 3 in a double logarithmic plotting. Since it is well known that the volume and the dry weight in a yeast cell increase with the degree of ploidy [7] a series of different polyplonts may be regarded as consecutive growth phases of one (artifically composed) organism. In this case only the maximum heat production and the corresponding weight for each polyploid culture were chosen for calculation. The results of our experiments as well as some other data which were calculated from the literature are listed in the table. For all organisms

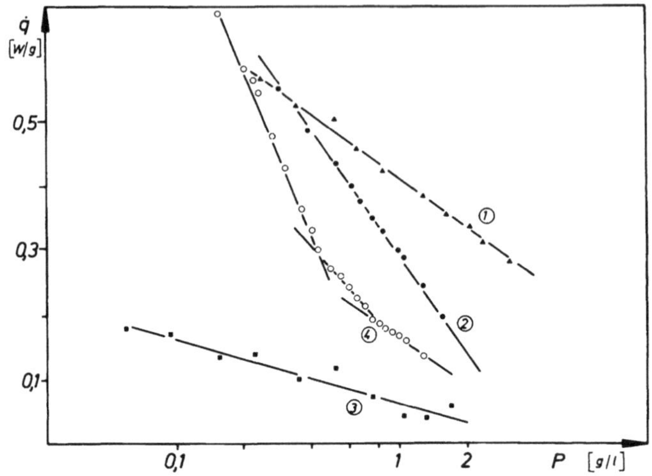

Fig.2: The specific heat production $\dot{q} = \frac{dq}{dt}$ as a function of the logarithm of the dry weight P, due to eq.(8). ①S.cerevisiae at 30°C, ②S.cerevisiae at 40°C, ③B.stearothermophilus, ④P.mirabilis.

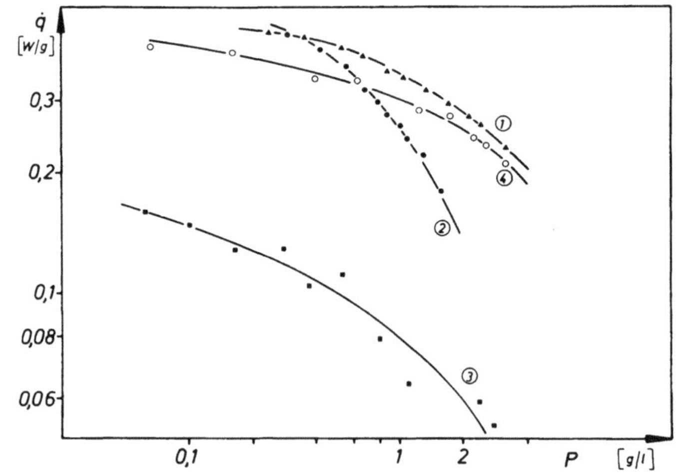

Fig.3: The logarithm of the specific heat production $\dot{q}$ as a function of the logarithm of the dry weight P. Plotting due to eq. (9). ①②③as in Fig.2, ④S.cerevisiae at 25°C.

organism	transition coefficient r [mW/g]	correlation factor due to eq.(8)	eq.(9)	author
Saccharomyces cerevisiae				
diploid 20°C	24	-0.97	-0.97	
25°C	32	-0.99	-0.97	
30°C	53	-0.99	-0.96	
35°C	61	-0.99	-0.97	
40°C	102	-0.99	-0.96	
haploid 30°C	41	-0.98	-0.96	
series of different ploidies	21	-0.98	-0.91	
Bacillus stearothermophilus	16	-0.96	-0.93	
Streptococcus faecalis **	204	-0.98	-0.98	[10]
Proteus mirabilis ** (different strains)	139	-0.99	-0.97	[9]
	71	1.00	-0.97	
	23	1.00	-0.97	
Frog	0.20 *			[3]
Fowl	1.35 *			[3]
Fish	0.23 *			[11]
Mouse**	1.90 *			[12]
Mouse**	4.70 *			[13]

Table: Transition coefficient r from equation (6) and correlation factors of a linear regression analysis due to a semilogarithmic plotting of equation (8) and a logarithmic plotting of equation(9).

* = calculated per wet weight

**= calculated from the data of the cited authors

the absolute values of the correlation factors for plotting due to equation (8) are closer to 1 than those due to (9). From this a better verification of equation (8) is noted. Furthermore it is evident from Fig.3 that plotting due to equation (9) always gives bent curves contrary to the semilogarithmic plotting of Fig.2 due to (8). In some cases we were unable to calculate the values over an appropriate part of the thermogram [8], [9], because of the insufficient precision of the reproduced diagrams in the literature. The data from the literature cited in the table for higher organisms are calculated per g of wet weight and not per dry weight as for microorganisms. Assuming a water content of about 90% all data will reach the same order of magnitude.

Discussion

From the correlation factors in the table adequate verification of equation (8) is evident. But on the other hand a wide discrepancy between the r-values for the different organisms is shown. Moreover, with the same organism, but at different temperatures or during a multiphasic growth (Fig.1, ③ and Fig.2 ④) different r-values are also obtained indicating diverse ways of metabolism. The r-value was expected to be an universal constant [1]. As this is not confirmed by the experimental results the variation may be explained by the entropy change $\left(\frac{\partial S}{\partial \xi}\right)$ due to the completness of reaction ξ , which was neglected in equation (1) [2]. Since ξ depends on the number of moles of reactants which may not necessarily remain constant in the course of reactions, it causes an entropy change and consequently a variable r-value. As microorganisms are known to work far away from thermodynamic equilibrium and to exhibit an active metabolism, the change of molecules is relatively high and is exposed to numerous outer parameters. Therefore, the differences between the r-values characterize to some extent the efficiency of metabolism, and r itself may represent a significant factor in comparative physiology.

Further experiments by direct calorimetry are necessary for clarification and understanding.

Acknowledgement: One of the calorimeters was placed to our disposal by the Deutsche Forschungsgemeinschaft.

References:

1. Zotin, A.I.: Thermodynamic aspects of developmental biology. S. Karger, Basel (1972).
2. Prigogine, I.: Introduction to thermodynamics of irreversible processes. Thournes, Chicago (1955).
3. Zotin, A.I., Grudnitzky, V.A.: Ratio between the heat production and respiration during animal growth. (Russian). Ontogenesis 1, 437-444 (1970).
4. Laskowski, W.: Inaktivierungsversuche mit homozygoten Hefestämmen verschiedenen Ploidiegrades. Zeitschrift für Naturforschung 15b, 495-506 (1960).
5. Lamprecht, I., Schaarschmidt, B., Stein, W.: Mikrokalorimetrische Untersuchungen zum Stoffwechsel von Hefen. III. Betriebsstoffwechsel in Glukosepuffer. Biophysik 10, 177-186 (1973).
6. Schaarschmidt, B., Lamprecht, I.: UV-irradiation and measuring of the optical density of microorganisms in a microcalorimeter. Experientia 29, 505-506 (1973).
7. Lamprecht, I., Schaarschmidt, B., Welge, G.: Microcalorimetric Investigation of the Metabolism of Yeasts. V. Influence of Ploidy on Growth and Metabolism. Rad. and Environm. Biophys. 13, 57-61 (1976).

8. Eriksson, R., Wadsö, I.: Design and testing of a flow microcalorimeter for studies of aerobic bacterial growth. In: Proceedings of the 1. European Biophysics Congress. Vol.IV, pp. 319-327. Edited by E. Broda, A. Locker and H. Springer-Lederer. Verlag der Wiener Medizinischen Akademie, Vienna (1971).

9. Jaworski, A., Sedlaczek, L., Czerniawski, E. and Zabłocki, B.: Thermogenesis of the stable L-forms of Proteus mirabilis and their parent bacteria. Acta Microbiologica Polonica 17, 219-230 (1968).

10. Forrest, W.W., Walker, M.G.: Thermodynamics of bacterial growth. Nature 196, 990-991 (1962).

11. Grudnitzky, V.A.: The dependance of heat production from body weight during the growth of Xiphophorus helleri (Russian). Doklady Akademia Nauk SSSR 202, 1435-1437 (1972).

12. Prat, H.: Calorimetry of higher organisms. In: Biochemical microcalorimetry, pp. 181-198. Edited by H.D. Brown, Academic Press, New York (1969).

13. Walker, M.G.: Heat production of the albino mouse during growth. Experientia 23, 541 (1967).

3.5 Calorimetric Studies of Lactic Acid Bacteria and the Effect of 2,4-Dinitrophenol on their Catabolic Regulation

P. Monk, W. Forrest, I. Wadsö

INTRODUCTION

It has been amply demonstrated that the ability of non-proliferating suspensions of many lactic acid bacteria to degrade glucose is a well defined property (1,2,3). Thus suspensions prepared in the same way may reliably be expected to have kinetic properties reproducible to better than 2 % (2), more stable than most enzyme preparations.

Oxygen may modify the fermentation of lactic acid bacteria and large heat effects have been observed when Streptococci grow in the presence of oxygen causing their thermograms to become complex (4).

Metabolic inhibitors and uncouplers have extensive influence on glycolysis in organisms carrying out oxidative phosphorylation. DNP is an uncoupler that is widely employed and its mode of action in this area is that of uncoupling catabolism from oxidative phosphorylation. It is also generally considered to affect only oxidative phosphorylation and not anaerobic glycolysis (5).

The relationship between four Streptococci and oxygen has been examined calorimetrically and in one case correlated with product formation. The effect of oxygen and DNP on catabolic regulation in nongrowing and growing cells has been investigated.

---

*On leave from the Australian Wine Research Institute

†The Australian Wine Research Institute, Private Bag, Glen Osmond, 5064 South Aust. Australia.

EXPERIMENTAL

Calorimetric experiments were carried out with a heat conduction flow microcalorimeter fitted with a flow through cell (6). Experimental details were the same as reported elsewhere (4).

RESULTS

Experiments with Washed Suspensions without DNP

When a washed suspension of Streptococci in phosphate buffer is pumped through a flow calorimeter a steady state baseline is established slightly above that for buffer alone. This heat effect is a measure of the small endogenous metabolism of the cells (7). On addition of glucose to the bacterial suspension being pumped to the calorimeter, the heat effect increases to a new steady state which includes the liberation of heat from catabolism of the glucose and an oxidative heat effect. On stopping the flow the established steady state is maintained for a short time, the heat effect then slowly decreases linearly and then exponentially to establish a new steady state of anaerobic glycolysis. Re-starting the flow causes the heat effect to increase to the initial steady state.

The bacterial suspension brings about the anaerobic state by utilization of the dissolved oxygen which was previously supplied by the flowing suspension. The time taken to establish the new steady state is dependent upon the bacterial cell density of the suspension located in the calorimeter cell and the oxygen status of the suspension at the time the flow was stopped.

A relationship can be established between the anaerobic (stopped flow) and aerobic (flow) heat effects for Streptococci, catabolizing glucose. By plotting the aerobic heat effect against the anaerobic heat effect for each organism a linear relationship is obtained. This relationship is reproducible for each strain and found to be independent of decreased ability to catabolize glucose when examined up to thirty hours after harvesting the cells. Different activities are obtained by varying the cell density of the suspensions.

The anaerobic heat effect is a quantitative measure of the rate of glucose degradation to lactic acid in a homolactic fermentation (2). The aerobic heat effect is the sum of the oxidative component of the suspension and the decreased anaerobic heat effect due to the oxidation of a fraction of a metabolic intermediate.

The reproducibility of the measurements demonstrates the ability of the technique to measure the oxidative ability of Streptococci and to differentiate between them on this basis.

The intermediate undergoing oxidation can be determined as pyruvate by catabolism in the presence of semicarbazide and degradation of added pyruvate.

Experiments with Washed Suspensions with 2,4-Dinitrophenol

When the molar turnover of glucose is measured anaerobically and aerobically no difference is observed in the rate of degradation, but the amount of lactate produced aerobically differs from a homolactic fermentation. This is due to the aerobic oxidation of pyruvate. The deviation from a homolactic fermentation is variable depending upon the organism.

If DNP is added to a suspension catabolizing glucose the rate of glycolysis increases and if the system is anaerobic the increased turnover of glucose appears as lactate and the fermentation is homolactic. The increased heat effect during stimulation can be correlated with increased lactate production.

When DNP is added over a range of 0-400 µmole/l to a suspension catabolizing glucose, the heat production both anaerobically and aerobically increases to a maximum rate at about 200 µmole/l. The increased heat production can be as high as 70 % of the basal rate.

Examination of the rate of glucose degradation immediately after the addition of glucose to a suspension with and without DNP, shows cells without DNP to regulate their catabolic activity to a lower rate within a few minutes after glucose addition. Cells with DNP however continue at the initial elevated rate.

GROWTH IN THE PRESENCE OF 2,4-DINITROPHENOL

Streptococci grown in a complex medium containing limiting glucose, with and without DNP, exhibit the same specific growth rate, but the cell yield is lower in the presence of DNP, possibly due to catabolic regulation being affected and glucose being degraded faster than the energy generated can be coupled to anabolic processes.

DISCUSSION

It is well established that different strains of Streptococci are micro-aerophilic and the strains exhibit differences in their ratio of oxidative to anaerobic heat effect. The precision with which this ratio can be measured suggests that this could be a useful taxonomic feature, being an extension of the suggestion already made that glycolytic activity should be used in chemical taxonomy of Streptococci (3).

Oxygen does not effect the rate of glycolysis, but does change the balance of products formed. In one experiment glucose gave 1.67 moles of lactic acid instead of two moles expected from a homolactic fermentation.

It is likely that the oxidation of pyruvate would give rise to acetate (8,9) which is supported by results of the following thermochemical calculations.

It is assumed that the steady state process under aerobic conditions can be described by

(1) Glucose + x O_2 $\rightarrow$ (2-x) lactate + x (acetate + carbonate)

The anaerobic state exists when $x = 0$.

Taking into consideration pKa values and enthalpies of ionization (10), the enthalpy change for reaction (1) was calculated from literature data (11) to be ΔH(phosphate buffer, pH = 6.2) = $-(117 + 509x)$ kJ/mole glucose. For the anaerobic condition ΔH will be -117 kJ/mole glucose. As the rate of glucose degradation is not affected by the presence of oxygen and the ratio between the heat effect for aerobic and anaerobic conditions, R, is thus given by

$$R = \frac{117 + 509x}{117}$$

For the culture studied we observed this value to be R = 2.35 leading to a value for x = 0.31 mole. Experimentally we found that 2 - x = 1.67 or x = 0.33 mole.

Cells catabolizing glucose in the absence of DNP are observed to effect a transition from a rapid initial rate to a lower rate. The transition is rapid, and has the characteristics of positive feedback so that regulation of glycolysis in this uncoupled condition appears to consist of the organism glycolyzing at a minimum rate. In the presence of DNP the transition does not occur and the cells retain their initial rapid rate of catabolism; the DNP inactivates the normal feedback regulation. The catabolism of glucose is already completely uncoupled in the washed non-growing suspension and the observed effect of DNP, normally considered as an uncoupler (5), obviously cannot be explained in terms of uncoupling.

In growth the addition of DNP was found to increase the rate of glycolysis similarly to the effect of non-growing suspensions, but the specific growth rate of the organism was not affected. The rate of synthesis of new biomass by each existing unit of biomass is thus unchanged so that the same amount of biologically available energy is coupled to synthesis either in the presence or absence of DNP.

The effect of DNP cannot be localised more precisely from the present experiments than to an interaction occurring before the production of endogenous pyruvate in the Embden-Meyerhof pathway, since its addition increases only the rate of glycolysis and lactate production; aerobic oxidation of pyruvate is unaffected by DNP. The maximum effect of DNP is a stimulation of anaerobic glycolysis by a factor of about 1.6. This is closely the same as the increased rate of glycolysis found during transient states in physiologically normal cells (12). The great difference in the effect of DNP is its concentration dependence, with a progressive increase to a maximum as the concentration is raised, compared with discontinuous changes observed in physiologically normal cells. It appears then that DNP is progressively removing the feedback control which reduces the rate of glycolysis to a minimum level, while having no other effect on the products of metabolism. Calorimetric observations are consistent with the hypothesis that no metabolic changes occur, except for increased lactate production.

A more detailed discussion will be reported elsewhere.

ACKNOWLEDGEMENTS

This work has been supported by the Swedish Board for Technical Development and the Australian Research Grants Committee.

REFERENCES

1. Boivenet, P.: These, Universite-d-'Aix-Marseille (1963).
2. Forrest, W.W., Walker, D.J., Hopgood, M.F.: Enthalpy changes associated with the lactic fermentation of glucose, J. Bact. 82, 685-690, (1961).
3. Manderson, G.J.: Ph.D. Thesis, Univ. Qld., Australia (1973).
4. Monk, P., Wadsö, I.: The use of microcalorimetry for bacterial characterization, J. Appl. Bact. 38, 71-74, (1975).
5. Lehninger, A.L.: Biochemistry, 2nd edn, p. 520, Worth Publishers, New York.
6. Monk, P., Wadsö, I.: A flow micro reaction calorimeter, Acta Chem. Scand. 22, 1842-1852, (1968).
7. Forrest, W.W., Walker, D.J.: Calorimetric measurement of energy of maintenance of Streptococcus faecalis, Biochem. Biophys. Res. Comm. 13, 217-222, (1963).

8. Dolin, M.I.: The DPNH-Oxidizing Enzymes of Streptococcus faecalis, Arch. Biochem. Biophys., 55, 415-435, (1955).

9. Demko, G.M., Blanton, S.J.B., Benoit, R.E.: Heterofermentative carbohydrate metabolism of lactose-impaired mutants of Streptococcus lactis, J. Bact. 112, 1335-1345, (1972).

10. Izatt, R.M., Christensen, J.J.: in Handbook of Biochemistry (Sober, H.A., ed.), pp. J-48-J-139. The Chemical Rubber Company, Cleveland, (1968).

11. Wilhoit, R.C.: in Biochemical Microcalorimetry (Brown, H.D., ed.), pp. 305-317, Academic Press, New York and London, (1969).

12. Forrest, W.W., Berger, R.L.: in First European Biophysics Congress IX (Broda, E. ed.), 304-305, (1971).

4. Calorimetry of Organs

4.1 Calorimetric Investigations on Animal Suborganisms: Organites, Tissues and Isolated Organs

P. Boivinet

INTRODUCTION

If we read the excellent reviews on biocalorimetry by C. Spink and I. Wadsö (1), I. Lamprecht (2) or B. Schaarschmidt and I. Lamprecht (3) it is apparent that, although the publications lie in the principal field of interest of the author, the list of references nevertheless gives an indication of the important fields of research: fifty per cent of the works cited are concerned with biochemical reactions, i.e. the molecular level of organisation, and thirty per cent with unicellular systems, microorganisms or blood cells (Wadsö at this meeting) whilst the remainder deal with subcellular systems, tissues, isolated organs and even whole pluricellular organisms. The reason for this uneven distribution may well be that it is possible to obtain well defined states in biochemical systems and, sometimes, to treat them thermodynamically. In the case of naturally isolated cells it is easy to obtain large populations, thus eliminating individual peculiarities. In both cases, homogeneous or almost homogeneous and isotropic systems are obtained, open to a wide variety of experimental techniques. On the other hand, complex and macroheterogeneous subsystems are difficult to investigate and often unpredictable. There is, however, an exception: for over fifty years now, excitable tissues (4 - 5), and particularly skeletal muscle, have been investigated with remarkable results by A.V. Hill, his coworkers and subsequent workers. This is the subject of a special paper at this meeting by Woledge. I can only point out that it was for this work that the only Nobel prize in microcalorimetry was awarded.

SUBCELLULAR SYSTEMS OR ORGANITES

Microcalorimetric research on subcellular systems has received rather little attention in the literature. In spite of the progress in differential centrifugation and electron microscopy it is still difficult to get sufficient quantities of homogeneous uncontaminated particles suitable for microcalorimetry. Furthermore these subcellular fractions are delicate and the "ecologic" conditions necessary to ensure a survival time compatible with the "lag phases" of the microcalorimeters are still not known.

Recent advances, however, in instrumentation of flow calorimeter (6 - 7) or highly sensitive apparatus with very short time constants, for example the NBS batch microcalorimeter (8 - 9) renew possibilities of progress in this field.

Two kinds of subcellular systems have been most frequently studied, membranes and mitochondria. Of the numerous experimental techniques reviewed by Levine (10) in studying the protido-lipidic structures we shall concern ourselves first with differential scanning calorimetry. This method is, indeed, the most straightforward in obtaining information on thermotropic transitions in protein-lipid complexes. These works on biomembrane phase transitions and related models have been reviewed recently by Chapman, Urbina and Keough (11). "Princeps" experiments (12 - 13) were carried out on microorganism cells (Mycoplasma laidlawii). It was possible by comparing living cells, isolated and purified membranes and their lipidic extracts to confirm the hypothesis of bilayer structure. Nowadays, the tendency is to concentrate more on artificial models or liposome dispersions (14 - 15). Suurkuusk et al (16) have reported on calorimetric and fluorescent probe studies of the gel-liquid cristalline phase transition in dipalmitoylphosphatidylcholine vesicles.

Papahadjopoulos et al (17) have studied the fluidizing effects of local anesthetics such as dibucaine on the solid to liquid phase transition in phospholipid liposomes. Cater et al (18) used a lecithin-water model to study the interactions of several drugs structurally related to morphine or similar to desipramine. There are examples where thermal methods promise to contribute to pharmacological interpretations. We are more concerned, however, with animal samples: Jackson et al. (19) using differential heat capacity calorimetry detected a thermal transition for human erythrocyte ghosts. This transformation is completely irreversible, probably because of the thermal denaturation of membrane protein. In agreement with this hypothesis, the ATPase attached to the membrane loses its activity in the same range of temperature.

Another type of investigations concerning biomembrane involves surface interactions. Kemp, for example (20), using an LKB batch calorimeter, determined the hydrolysis of exogenous ATP on muscle fibroblasts isolated with trypsin.

The inhibition of the enzyme by ouabain, oligomycin, mersalyl and sulfhydryl blocking agents was studied. The fixation on the membrane of rabbit antifibroblast serum was also reported. Similarly Zala et al.(21) have reported on the enthalpy change associated with the binding of glucose to human red blood cell ghosts.

The pioneers of mitochondrial microcalorimetry are Poe, Gutfreund and Estabrook (22 - 23). Studies of the ΔH are particularly suitable in determining the energy balance in oxidizing processes.

Their initial work was concerned with ETP (electron transport particles) isolated from the heart. Subsequently they studied the energy coupling between catabolism and the absorption of calcium (24) or potassium (25) in the mitochondria of rat liver. An oxygen electrode in the calorimeter enables a comparison to be made between direct and indirect calorimetric methods. The results depend with the type of anions surrounding the mitochondria. The enthalpies of oxidation of the substrate were

- 44,1 kcal/atom oxygen with chloride as principal anion,
- 35,8 kcal/atom oxygen with acetate and
- 34,0 with phosphate.

These differences were attributable to the heat of proton neutralization which varies with the anionic environment. The enthalpy of calcium binding, - 25 $\pm$ 0,85 kcal/at. was determined from the energy balance. Its value indicates that some of the calcium is not in free solution.

Tyng-Fang Chien and Burkhard (26) explain the discrepancy observed by Poe et al. when they compared the enthalpy of oxidation of succinate by oxygen with that by ferricyanide as being due to the terminal electron acceptor. The values obtained for oxygen with non-phosphorylating submitochondrial particles or dinitrophenol-uncoupled mitochondria agree with those calculated thermodynamically.

On the other hand, the enthalpy of oxidation by ferricyanide measured for ultrasonic dispersed mitochondria agreed with the calculated value, but those for uncoupled mitochondria did not. If the release of the protons is taken into account very good agreement between calculated and experimental results are obtained in every case.

As for subcellular particles other than membranes and mitochondria, ribosomes have been investigated by Lamprecht et al. (27) but only on a cell free protein synthetizing system of Saccharomyces and hence not on an animal system.

TISSUES AND ISOLATED ORGANS

Here three kinds of difficulties arise:

a) - ensuring sufficient thermal contact between sample and the calorimeter wall.

b) - maintaining adequate oxygenation, correct supply of metabolites and efficient removal of waste products with compacted mass of cells. Perifusion is, however, not as efficient as natural perfusion. It is possible to obtain a homogeneous system by isolating the tissue cells with trypsin or collagenase, if this solution is compatible with the physiological behaviour of the sample.

c) - defining precisely the base line (experimental zero). Living samples are continuously metabolizing and an initial record of the base line at the beginning of the experiment is not sufficient.

Wadsö's drop calorimeter devices minimize this inconvenience, but generally it is necessary to kill the tissues by injecting a suitable toxic material and determine the terminal reference base line. Contamination by microorganisms also causes a shift in the base line.

Hence, an antibiotic which does not interfere with metabolic activity is needed.

After Woledge's splendid lecture, it is obviously not possible to add anything on the subject of muscles. Historically, so far as the principal authors are concerned, the investigations on heat production in nerves were directly comparable with those on muscle. Helmholtz in 1848, using thermocouples, was the first to attempt to record the heat production in activated nerves, but without success.

About three quarters of a century later, Hill succeeded in making a measurement (28). Abbot et al. (29) have shown that the initial heat produced in crab limb nerves is diphasic in character. An exothermic phase lasting about 60 msec, with a mean value of 14. 10^{-6} cal/g (wet weight) coincides with an increase in the action potential (A. P.). An endothermic phase lasting about 300 msec follows, with a mean value of 12. 10^{-6} cal/g (wet weight), which coincides with the decrease in the A.P.

The origin of the initial heat has been ascribed to the Joule and Peltier effects in the heterogeneous tissues and to ionic interchange across the membrane. Energy capacity stored in the membrane was released during the increase of A.P. and restored during the decrease of A.P. Structural changes, i.e. entropy changes in the membrane, may also occur. The initial heat is followed by the heat of recovery, as in the muscle.

The two phase character has been observed as a general feature in all forms of nerves. Ritchie (30), in a review on the energetics of nerve conduction, maintains that the giant squid axon was the best for electrophysiology because of its large diameter (500 µm), but that the non myelinated fibres of the cervical vagus nerves of the rabbit are better suited for determining heat exchanges. These fibres, diameter 0,4 - 1,2 µm, form 6000 cm^2 of membrane per g wet weight, compared with 80 cm^2 for the giant axon, and the heat produced is related to the surface-to-volume ratio. Abbot (31), Aubert and Keynes (32-33) studied heat production in the electric organs of certain species of fishes (*Torpedo*, *Electrophorus*, *Malapterus*). As for nerves, an initial two-phase heat production was recorded, but the exothermic effects appear to be due to an artifactual Joule effect in damaged tissues. The cooling (endothermic) phase is the more significant event and is followed by a stage of classical metabolic heat recovery. Thermal inertia of the system prevents any correlation with the phases of the A.P. The Peltier effect heat of ionic mixing and structural changes of the membrane seem to have been discarded as possible explanations.

The reversibility of the sodium pump appears to be the principal source of heat.

The last example of heat production in an excitable tissue is by Hoffert et al. (34), who have shown by direct and indirect calorimetry that aerobic processes provide 40 - 50 per cent of the total energy in the isolated retina of teleost rainbow trout (*Salmo gairdneri*), the remainder coming from anaerobic sources. These results show the obvious disadvantage of exclusive use of respirometric techniques, and also the advantages offered by direct calorimetry.

FAT TISSUES

Storage and release of energy and their hormonal regulation are fields of much interest in physiology. Chinet and Girardier (35), using a thermal flux differential calorimeter measured the calorigenic response of rat brown fat tissue to norepinephrine. They have shown the unexpected potentiation of the response by inhibitors of hormone binding, i.e. DOPA or vitamin C. Their results suggest that the inhibitors act as competitor for the sites at which the hormone either has an inhibiting effect on calorigenesis or where norepinephrine is destroyed. Chinet et al. (36) have likewise measured the energy flux associated with sodium removal in the brown fat tissue of the rat and compared it with that for rat solear muscle.

The blocking effect of ouabain on the sodium pump and the measurement of the sodium flow with isotope 22 of Na have shown a 50 % yield in sodium pumping when using ATP.

Insulin greatly increases the glucose absorption rate of several tissues: muscles such as diaphragm, but mainly fat cells (adipocytes). Boivinet et al. (37) measured the effect of insulin on rat epididymal fat tissue, using a Calvet calorimeter.

Wistar male rats thirty and ninety days old, were quickly killed without anaesthesia. About 100 mg of epididymal fat was removed aseptically and placed in the calorimeter in 1.2 ml of Krebs-Ringer solution, and 4 g/l of glucose and 2 g/l gelatin were added. Gelatin prevented adsorption of insulin on the glass walls (Fig.1).

The solution was saturated with 95 % oxygen and 5 % CO_2 (carbogene). This gas also filled the free space in the cell (ca 12 ml). Penicillin was used to prevent bacterial contamination (3000 U.I/ml). The insulin solution and the metabolic inhibitor (eg. NaF) were successively injected by means of syringes placed in the calorimeter tube, insulin after the heat without insulin had been recorded, NaF at the end of the experiment (Fig.2).

The fraction $(\Delta_i - \Delta_m)/\Delta_m$ obtained from the thermogram (Fig.2) gives a relative measure of the insulin activity. In the concentration interval of 25 to 110 µU of insulin/ml the fraction is propor-

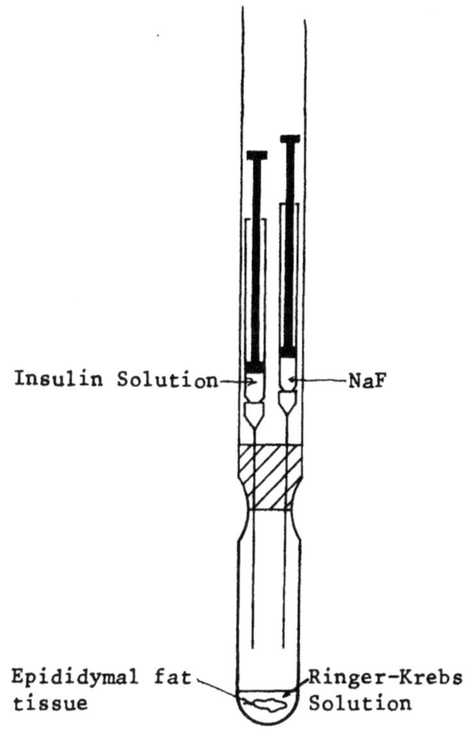

Fig. 1 : Microcalorimetric determination of insulin activity. About 100 mg of epididymal fat tissue and 1.2 ml of Krebs-Ringer solution are put in the calorimetric cell. Two syringes make possible the injection of insulin solution and, at the end of the experiment, of NaF solution which determines the final reference base-line.

tional to the quantity of insulin injected. The accuracy obtained is about 15 %. Gulfe and Hellmann (38) studied the heat production of pancreatic ß cells stimulated by glucose in islets of Langerhans obtained from obese hyperglycemic mice. The heat production, measured in the LKB batch microcalorimeter, was compared with microrespirometric or micropolarographic oxygen determinations. Islets were isolated

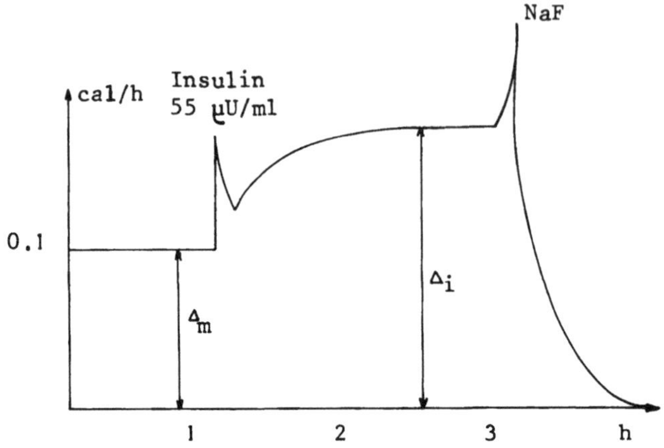

Fig. 2 : Thermogram of determination of insulin activity
Δ_m : Metabolic activity without insulin
Δ_i : Metabolic activity in presence of insulin

by a collagenase treatment. Tylosin or gentamycin was used to prevent bacterial growth. The degree of heat production in the absence of glucose was equivalent to about 50 nW/islet or 13 mW/g islet dry weight. After mixing, increasing the glucose concentration to 20 mM, the heat production increased to about 90 nW/islet. The antimycin suppressed all heat production before as well as after the addition of glucose. It is interesting to note that the production of heat increased 80 % with reference to the base level after exposing the islets to 20 mM glucose, whereas oxygen absorption increased by only 50 %.

This suggests an increase in the calorific coefficient of oxygen with glucose concentration when the ß cells use correspondingly more carbohydrate as a source of energy. Schaarschmidt and Schaefer (39) have given details of preliminary microcalorimetric experiments on human cutaneous cells (epidermis) and on guinea-pig liver tissue.

The values obtained without substrate in the buffer (endogenous heat production) and with 10 g/l glucose were compared with the heat production of yeast, in air, in N_2 (2 kp/cm^2) and in O_2 (1 kp/cm^2).

Microcalorimetry would appear to be promising in research into the metabolism of cancerous cells but, as yet, there is only one investigation which can be quoted, i.e. that of Kresheck and Gordon (40) who gave some preliminary results on cell cultures.

ISOLATED ORGANS

The heart is a good example of an isolated organ and, at present, the most investigated.

A study of the energy balance of the isolated heart requires, firstly, a good knowledge of the mechanical working, and also of the total energy flux produced by aerobic and anaerobic metabolic means.

So far there has been only one work reported on a whole isolated mammalian heart, and that is rather old. Neill and Huckabee (41) used two dogs for each experiment. The removed heart was continuously supplied with blood by the other dog (blood donor dog). The isolated heart was kept in a thermally insulated box. The heat production was calculated from the blood flow and from the temperature difference between the blood entering and leaving the heart. The oxygen consumption was measured by manometric technique. An intraventricular ball converted the mechanical work into heat. The results, a mean of seven determinations, showed an anaerobic heat production corresponding to 12,5 % of the total energy requirement. This is rather surprising for a higher animal heart, but calibration and measurement of the temperature and flow, and conversion of work into heat are very difficult. The Grayson's technique (43) of internal calorimetry, using heated thermocouple probes, "in vivo" is better, but still complex. Consequently Ricciuti et al.(42) preferred to revert to more established techniques developed by Fenn for the mechanical work of skeletal striated muscle and by Hill for thermal measurements. They therefore used cardiac trips and papillary muscles which can be assimilated in linear elements and hence studied using Hill thermocouples. In the rabbit capillary muscle the heat production rate varied between 80 and 140 mcal/min under experimental conditions.

In the case of lower animals hearts, it is easier to maintain the organ alive and consequently more work has been published. The earliest were by Bohnenkamp et al.(44) and Fisches (45) on poikilotherms (eel, tortoise, frog). Boivinet and Rybak (46) have carried out microcalorimetric studies of the frog heart when completely open and under extension. The technique of the open extended heart, which was developed originally by Rybak (47), improves the interchange with the surroundings. Consequently, mechanical catalysis (48) cardiac contractions may be prolonged over many hours without any perfusion.

The frog's heart, weighing about 50 mg, is removed, opened and extended under a known tension in an extensometer(49) and then fixed on a paraffin plate and placed in the calorimeter with 1 ml of Ringer solution (Fig.3). Two platinum wires embedded in the paraffin enabled the electrocardiogram (E.C.G.) and the number of cardiac cycles to be recorded during the microcalorimetric experiment. Fig.4 shows a typical thermogram. The final reference line was obtained using a blocking solution of KCl and NaF (0.5 M). Measurements were made in air (heat production rate 0.985 cal/h.dry weight)and in 75% O_2,25% N_2 (2.54 cal/ h·g dry weight).

The heat produced in a single contraction cycle can be calculated from the electrocardiogram:

$$6.5 \cdot 10^{-4} \text{ cal/cycle} \cdot \text{g at } 27^{\circ}\text{C in 75 \% oxygen,}$$

$$2.88 \cdot 10^{-4} \text{ cal/cycle} \cdot \text{g at } 27^{\circ}\text{C in the air.}$$

The results have confirmed that the frog heart was under normal oxygen pressure partially anaerobic. Polarographic determination of the oxygen (47) showed that a partial pressure of oxygen greater than 250 mm was necessary for perfect anaerobiosis. In the preceding experiments, the heart was in a gaseous environment. It is possible in liquid environment , which is more efficient for removal of waste products, to use the catalase activity of the heart for oxygenation (Fig.5). For this 1g/1 $NaBO_3$ is added to the Ringer solution. After graphical correction of the thermogram, it is possible to determine

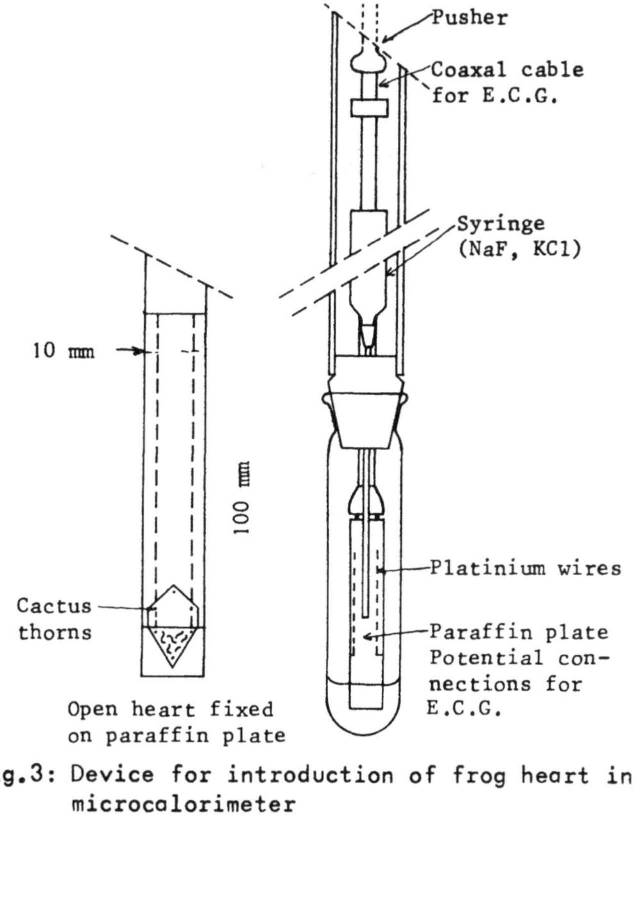

Fig.3: Device for introduction of frog heart in Calvet microcalorimeter

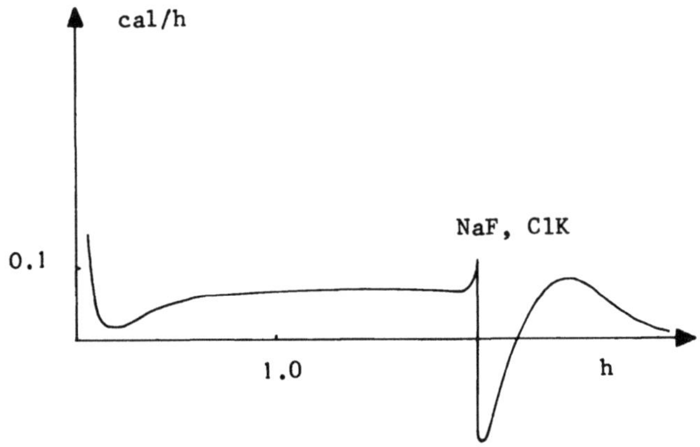

Fig.4: Thermogram of open extended heart in air at 27 °C

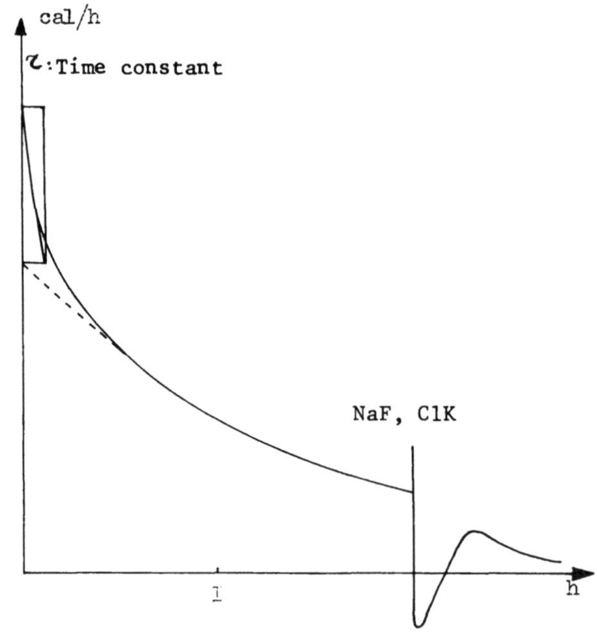

Fig.5: Thermogram of open extended heart with catalase oxygenation in liquid environment

the initial catalase activity of the myocard.

Recently, Herold (50,51) carried out an important and complete study of the energetics of myocard in the isolated ventricle of the snail (Helix pomatia). The ventricle work was first measured with a classical device (Fig.6). The isolated heart , fixed on a Straub cannula, became part of a recording system with constant pressure and its mechanical power P was measured. The ventricular heat production rate or total power $\Sigma(-\dot{\Delta}H)$ was determined using a Calvet microcalorimeter.

Two ventricles functioning isometrically under known load were stretched on a stainless steel frame .The predetermined load varies from 150 to 650 mg. The frame with attached hearts was placed in the microcalorimeter cell containing hemolymph or another suitable physiologic solution (Fig.7). A gas was bubbled through, a mixture of O_2 and N_2 or air, at a rate of 40 ml/h to maintain a constant P_{O_2}.

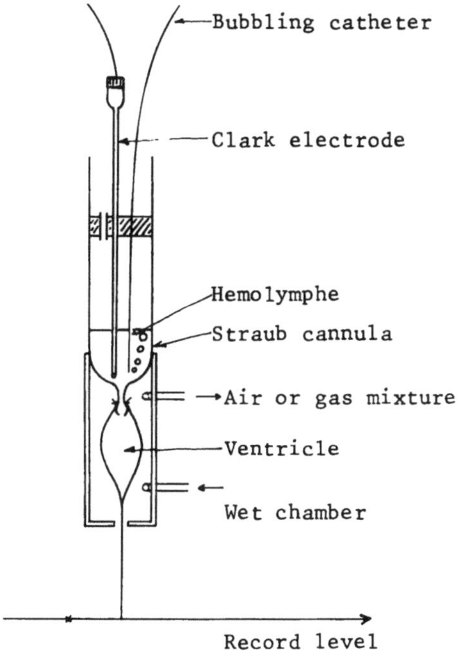

Fig.6: Snail heart, mechanical power measurement device

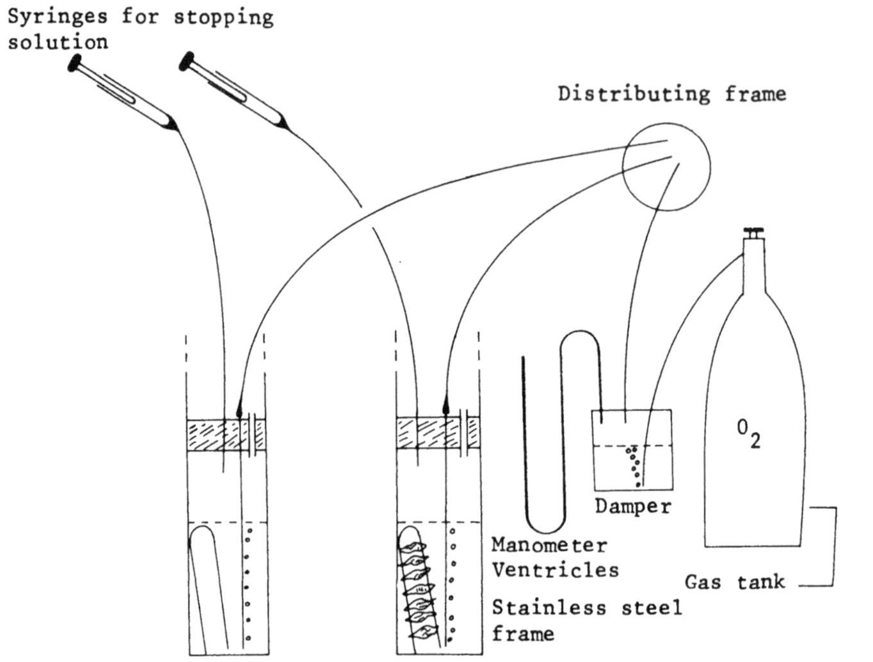

Fig.7: Snail heart, calorimetric device with differential oxygenation

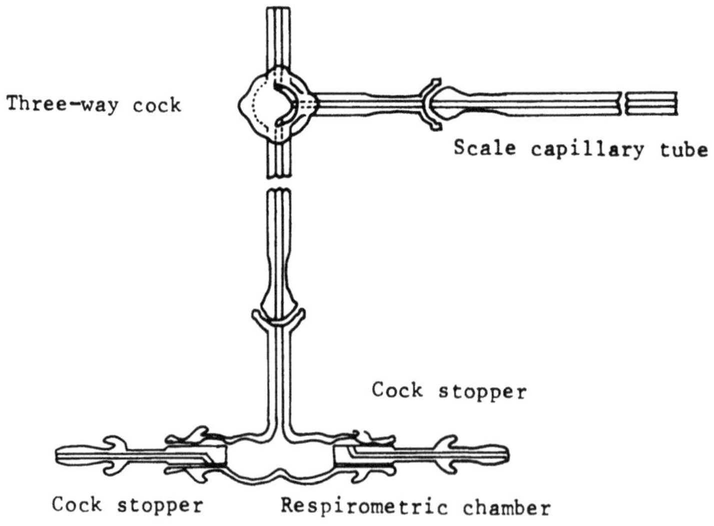

Fig.8: Snail heart, microrespirometric determination of O_2-uptake

The control cell contained the same volume of solution and the same volume of gas was bubbled through. Thus, the parasitic effect due to bubbling was eliminated. The base reference line, or experimental zero, was determined at the end of the experiment by injection of 10% formaline solution which eliminated all thermogenesis. It was thus possible to compare values of mechanical power with total power. The efficiency coefficient is defined after Gibbs as:

$$\alpha = \frac{\text{mechanical power}}{\text{total power}} = \frac{P}{\Sigma(-\Delta\dot{H})}$$

It varies with applied tension from 0.08 to 0.26 for an active isolated heart in contact with the hemolymph.

The comparison between mechanical power P and O_2-absorption was determined manometrically (Fig.8), or polarographically (Fig.9) and used to define the energy/oxidation coefficient ß

$$\beta = \frac{\text{mechanical power}}{\text{extra oxygen absorption}} = \frac{P}{O_2}$$

(extra absorption = active absorption - rest absorption)

ß varies between 0.20 and 1.08, depending on experimental conditions. The efficiency coefficient α and the energy/oxidation coefficient ß give a good indication of the energy characteristic of an organ, allowing comparisons with other systems.

Thus in the snail's heart, the amplitude of variations of ß is larger than in vertebrate hearts. It shows the very good adaptation of snail's ventricle to variable load and to varying oxygenation. Like the frog's heart, snail's heart under physiological conditions, is essentially anaerobic. The thermogenesis of ventricle may reach 1 cal/h·g wet weight or 40 µW/heart during maximum activity.

The factors influencing the energetic behaviour were systematically

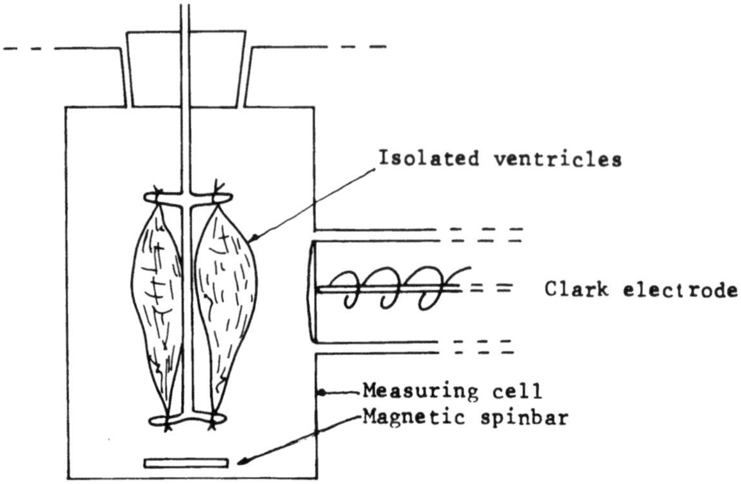

Fig.9: Snail heart, polarographic determination of O_2-uptake

studied, principally ionic equilibrium, viscosity of the perfusing liquid, active pharmacological substances, hormones and neurosecretions. The cardiotonic properties of hemolymph protein were demonstrated. The mechanical tension, however, appears to be the principal parameter of energy release.

GENERAL CONCLUSIONS

The strict oxygen dependence for the homeothermic neurones and the numerous experiments carried out on intact higher animals to establish the validity of the first law of thermodynamics in the living world have convinced many physiologists of the universality of indirect calorimetry, i.e. of respirometric techniques. The total molecular balance is often seemingly tedious. However, even in homeotherms some tissues live normally in the hypoxic state using anaerobic pathways: in the whole animal, this is concealed by cooperation between the organs. Thus the liver, in Cori's cycle, oxidizes the lactic acid produced by the muscle, and direct and indirect calorimetry will give the same value for total energy production. With isolated tissues such a presumption may not be valid. It is even more obvious in lower heterothermic animals. Direct calorimetry appears to be the most suitable and applicable means of detecting such phenomena and enables a more rigorous approach to be adopted, regarding energy yields and rectification for respirometric or analytic data. Calorimetry is, certainly, unspecific: heat appears, following the law of the entropy, as the universal "excretum". Thus recording heat is an adequate procedure for studying dissipative processes such as life, with the minimum of interference. In the case of tissues or isolated organs the natural, physiological conditions are generally reproduced: heterogeneity, darkness, constant temperature, no need of additional reagents, no agressive intervention such as radiation. It is often advantageous in detecting unsuspected features not to work under highly specific conditions.

REFERENCES

(1) Spink C. and Wadsö I. : Calorimetry as an Analytical Tool in Biochemistry and Biology, in "Methods in Biochemical Analysis" (editor D.Glick),Vol.23,pp. 1-159, Wiley-Science,New York (1976).

(2) Lamprecht I. : Biological Calorimetry and DTA, in "Ecole d'Eté de Calorimétrie", Recueil des conférences, AFCAT, 202-253, Paris (1973).

(3) Schaarschmidt B. and Lamprecht I. : Heat Production in Living Processes, in "Thermodynamic of Biological Processes" (editor A.I. Zotin) Nauka-Press, Moscow (in the press).

(4) Abbot B.C. and Howarth J.W. : Heat Studies in Excitable Tissues. Physiol. Rev., 53 (1), 120 - 158 (1973).

(5) Woledge R.C. : Heat Production and Chemical Change in Muscle, in "Progress in Biophysics and Molecular Biology", vol. 22, 39 - 74, Pergamon Press, N.Y. (1971).

(6) Monk P. and Wadsö I. : Flow Microcalorimetry as an Analytical Tool in Biochemistry and Related Areas. Acta Chem. Scand., 23, 29 - 36, (1969).

(7) Picker P., Jolicoeur C. and Desnoyers J.E. : Steady State and Composition Scanning, Differential Flow Microcalorimeters, J. Chem. Thermodyn., 1, 469 - 483, (1969).

(8) Prosen E.J. : National Bureau of Standard Report NBSIR - 73 - 179 Washington, DC (1973).

(9) Prosen E.J. and Goldberg R.N. - National Bureau of Standard Report NBSIR - 73 - 180 Washington, DC (1973).

(10) Levine Y.K. : Physical Studies of Membrane Structure in "Progress in Biophysical and Molecular Biology" (Editors : J.A. Butler and D. Noble) 24, 1 - 74., Pergamon Press (1972).

(11) Chapman D., Urbina J., and Keough K.M. : Biomembrane Phase Transitions. J. Biol. Chem., 249, 2512 - 2521, (1974).

(12) Hintz H.J. and Sturtevant J.M. : Calorimetric Studies of Dilute Aqueous suspensions of Bilayers formed from synthetic L-α-Lecithins. J. Biol. Chem., 247, 6071 - 6075 (1972).

(13) Reinert J.C. and Steim J.M. : Calorimetric Detection of a membrane - Lipid phase transition in living cells. Science, 168, 1850 - 1852(1970)

(14) Ververgaert P.H.J., Verkles A.J., Elbers P.F. and Vandeenen L.L. : Analysis of the cristallization process in lecithin liposomes : a freech etch study. Biochim.Biophys.Acta, 311 (3), 320 - 329 (1973)

(15) Klopfenstein W.E., de Kruyff B., Verkleij A.J., Demel R.A. and Van Deenen L.L.M. : Differential Scanning Calorimetry on Mixtures of Lecithin, Lysolecithin and cholesterol. Chemistry and Physics of Lipids, 13, 215 - 222 (1974).

(16) J. Suurkuusk, Lentz B.R., Barenholz Y., Biltonen R.L. and Thomson T.E. A calorimetric and fluorescent Probe Study of the Gel-Liquid Crystal.-line phase Transition in small single-lamellar Dipalmitoylphosphatidylcholine vesicles. Biochem., 15, 1393 - 1401 (1976).

(17) Papahadjopoulos D., Jacobson K.A., Poste G. and Sepherd G. : Effects of local Anesthetics on Membrane Properties. I. Changes in the fluidity of phospholipids bilayers. Biochem. Biophys. Acta, 394, 504 - 519, (1975)

(18) Cater B.R., Chapman D., Hawes S., and Saville J. : Lipid Phase Transition and Drug Interactions. Biochem. Biophys. Acta, 363, 54 - 69, (1974).

(19) Jackson W.M., Kostlya J., Nordin J. and Brandts : Calorimetric Study of Protein Transitions in Human Erythrocyte Ghosts. Biochem., 12, (19) 3662 - 3667, (1973).

(20) Kemp R.B. : Surface-localized Reactions in Tissues Cells : a microcalorimetric Study. Science Tools, 19, 3 - 5 (1972).

(21) Zala C.A., Jones M.N. and Levine M. : The Detection of the Enthalpy of binding of D. Glucose of Human Red Blood cellmembranes by Microcalorimetry. FEBS LETTERS, 48, (2), 196 - 199, (1974).

(22) Poe M., Gutfreund H. and Estabrook R.W. in "Methods in Enzymology" (Estabrook Pullman ed.), Vol. X, 708 - 714, Academic Press N.Y. (1967)

(23) Poe M., Gutfreund H. and Estabrook R.W. : Kinetic Studies of temperature changes and oxygen uptake in a differential calorimeter : the Heat of oxidation of NADH and succinate. Archiv. Bioch. Biophys., 122, 204 - 211, (1967).

(24) Poe M. : Kinetic Studies of Temperatures changes an oxygen uptake in a differential calorimeter : Energy Balance during Calcium Accumulation by Mitochondria, Archiv. Biochem. Biophys., 132, 377 - 387 (1969)

(25) Poe M. and Estabrook R.W. : Kinetic Studies of temperature changes, oxygen utilisation, pH and potassium concentration changes in a differential calorimeter : The enthalpy of potassium accumulation by Rat liver mitochondria. Arch. Biochem. Biophys., 128, 725 - 733 (1968).

(26) Tyng-Fang Chien and Burkhard R.K. : Stoichiometry and calculated Enthalpy change for Mitochondrial Oxidation of succinate by ferricyanide. J. Biol. Chem., 250, (2), 553 - 556 (1975).

(27) Lamprecht I., Lochmann E.R., Pietsch I., Uttech K. : Microcalorimetric Experiments on a cell Free Protein Synthetizing System of Saccharomyces, in "Protides of the Biological Fluids" (Editor : H. Peeters) Vol. 20, 551 - 553 - Oxford (1973).

(28) Hill A.V. : The Heat Production of Nerve. J. Pharmacol., 29, 161 - 165, (1926).

(29) Abbot B.C., Hill A.V. and Howarth J.W. : The positive and negative Heat Production associated with a Nerve Impulse. Proc. Roy. Soc. (London) Sec B, 148, 149 - 187 (1958).

(30) Ritchie J.M. : Energetic aspects of Nerve conduction : The Relation ships between Heat Production, Electrical Activity and Metabolism in "Progress in Biophysics and Molecular Biology" (Editor : J.A.V. Butler and D. Noble) Vol. 26, 147 - 187, Pergamon Press Oxford (1973).

(31) Abbot B.C.,Aubert X., and Fessard A. : La production de chaleur associée à la décharge du tissu électrique de la Torpille, J. Physiol. (Paris), 50, 99 - 102 (1958).

(32) Aubert X. and Keynes R.D. : The temperature changes during and after the discharge of the electric organ in Electrophorus electricus. Proc. Roy. Soc. (London) Ser B 169, 241 - 263, (1968).

(33) Keynes R.D. : The temperature changes during and after the discharge of the electric organ in Malapterus electricus. Proc. Roy.Soc (London) Ser B 169, 265 - 274 (1968)

(34) Huffert, Russel J., Craig E., and Paul O. : Comparison of direct and indirect calorimetry as a method for determining total energy produced by the isolated retina of rainbow trout (Salmo gairdneri) Exp. Eye. Res., 19, (4), 359 - 365, (1974).

(35) Chinet A. and Girardier L. : Potentiation of Brown Adipose Tissue Calorigenesis by Inhibitors of Hormon Binding. Experientia, 30 (6), 677, (1974).

(36) Chinet A., Clausen T. and Girardier L. : Mesure en calorimétrie directe du flux énergétique lié au transport de sodium. J. Physiol. (Paris), 71, (2), 271, (1975).

(37) Boivinet P., Garrigues J.C. and Grangetto A. : Dosage microcalorimétrique de l'insuline, Compt. Rend. Soc. Biol. France, 162, 1770 - 1774 (1968).

(38) Gylfe E. and Hellamn B. : The Heat Production of Pancreatic B cells stimulated by Glucose, Acta Physiol. scand., 93, 179 - 183, (1975).

(39) Schaarschmidt B and Schaefer : Stoffwechseleuntersuchungen an Hautund Lebergewebe, in : Lamprecht I. and Schaarschmidt B., Pressedienst Wissenschaft F U Berlin, 4, 74 - 75, (1974).

(40) Kresheck G.C. : The Heat produced in vitro with L 1210 cells, Cancer Biochem. Biophys., 1, (1), 39 - 42, (1974).

(41) Neill W.A. and Huckabee W.E. : Anaerobic Heat Production by the Heart J. of Clin. Invest., 45, 1412 - 1420, (1966).

(42) Gibbs C.L., Mommaerts W.F. and Ricchiuti N.V. : Energetics of cardiac contractions, J. Physiol. Lond., 191, 25 - 46, (1967).

(43) Grayson J., Coulson R. and Winchester B. : Internal calorimetry, assesment of myocardial blood flow and heat production, J. appl. Physiol., 30, (2), 251 - 257, (1971).

(44) Bohnenkamp H., Ernst H.W. and Nameto A. : On actual value of energy changes, its independance on mechanical work and its economy, Z. Biol. 88, 429 - 450, (1929).

(45) Fischer E. : The Heat liberated by the beating heart, Amer. J. Physiol. 96, 78 - 88, (1931).

(46) Boivinet P. and Rybak B. : Etude microcalorimétrique du fonctionnement électromécanique du coeur de grenouille, Life Sciences, 8, (II) 11 - 20, (1969).

(47) Fabel H., Lubbers D.W. and Rybak B. : Mesure polarographique de la consommation d'oxygène du coeur étalé ouvert de grenouille en contractions régulières prolongées "in vitro", Bull. Soc. Chim. Biol., 46, 811 - 817, (1964).

(48) Rybak B. : Nouvelles perspectives en physiologie cardiaque, Les Conférences du Palais de la Découverte fasc. 226, (1957)

(49) Rybak B. : Prolégomènes à l'extensométrie cardiaque expérimentale. C.R. Acad. Sc. Franc., 250, 3391 - 3393, (1960).

(50) Herold J.P. : Myocardial Efficiency in the isolated ventricle of the snail Helix pomatia L., Comp. Biochem. Physiol. 52 A, 435 - 440,(1975)

(51) Herold J. P. : De l'énergétique du ventricule isolé d'un mollusque, l'escargot, (Helix pomatia L.), Thèse, Besançon (France), (1975).

4.2 Calorimetric Studies of Muscle and Muscle Proteins

R.C. Woledge

It has been discovered during the past few years that much of the energy which is produced during contraction of skeletal muscle does not come from the splitting of ATP and the reactions which resynthetise it. It seems that this energy comes from changes in the proteins in the muscle, quite probably from changes in the contractile proteins. We would like to discover the identity of these hypothetical changes and are making two attacks on the problem. Firstly, in studies of living, excised muscle we investigate the magnitude of the unexplained part of the energy produced during contraction, and how it varies with different conditions. Secondly by studying purified contractile proteins in vitro we try to identify possible exothermic reactions which could be the source of the unexplained energy in the muscle.

The original work reported here has been done in collaboration with Dr Nancy A. Curtin, Dr Takao Kodama, Dr Howard White and Dr Ian Watson.

BIOCHEMISTRY OF MUSCLE

For the benefit of readers who may have no knowledge of muscle I will briefly summarise a few facts about muscle biochemistry which are needed for understanding this article. Interested readers who wish for further information are referred to the books by Carlson and Wilkie (1) and Wilkie (2).

Intact muscle contains two very active ATPase systems. One of these is the actomyosin system, consisting of the two contractile proteins actin and myosin, which are arranged in separate filaments in living muscle. ATP is split during the interaction between these filaments which is also responsible for the mechanical events of contraction. The reaction is controlled by the proteins troponin and tropomyosin which influence the actin in response to changes in intracellular Ca^{2+} levels. The Ca^{2+} level is itself controlled by the second ATPase, that of the sarcoplasmic reticulum, by which ATP splitting is coupled to the accumulation of Ca^{2+}. This calcium is released during stimulation, activates the contractile system, and is then reaccumulated within the sarcoplasmic reticulum.

In spite of the activity of these two ATPases, changes in the ATP concentration in muscle do not usually occur during contraction because the ATP is rapidly regenerated by the creatine kinase reaction:

$$\text{ADP} + \text{phosphocreatine} \longrightarrow \text{ATP} + \text{creatine}$$

If this system is exhausted or if the enzyme is blocked (by DNFB) then the adenylate kinase and adenylate deaminase system resynthesises the ATP:

$$2\text{ADP} \longrightarrow \text{AMP} + \text{ATP}$$

$$\text{AMP} \longrightarrow \text{inosine monophosphate} + NH_3$$

Muscle cells can of course also resynthesise ATP by glycolysis and by oxidative phosphorylation, but these processes are much slower and need not concern us here.

Myosin. The myosin molecule is large (MW = 500,000) and consists of a tail region carrying two heads. The tail is insoluble at low ionic strength and is responsible for the aggregation of myosin into filaments which occurs in vivo and can also be produced in vitro. The two heads, which are probably identical, can interact with actin and with ATP. For many purposes it is convenient to remove the tail of the myosin molecule (by treatment with a proteolytic enzyme) and to study only the heads, either singly (S_1) or in pairs joined by a short part of the tail (HMM).

Actin. This is a fairly small protein (MW = 40,000). The monomers (G-actin) have a strong tendency to form linear polymers (F-actin). It is

F-actin which exists in muscle and which is usually used for in-vitro experiments.

Actomyosin as an ATPase. From kinetic studies a mechanism has been proposed for actomyosin ATPase which can be simplified as the following cycle of reactions. (My represents myosin and Ac actin. The full arrows show the probable mechanism)

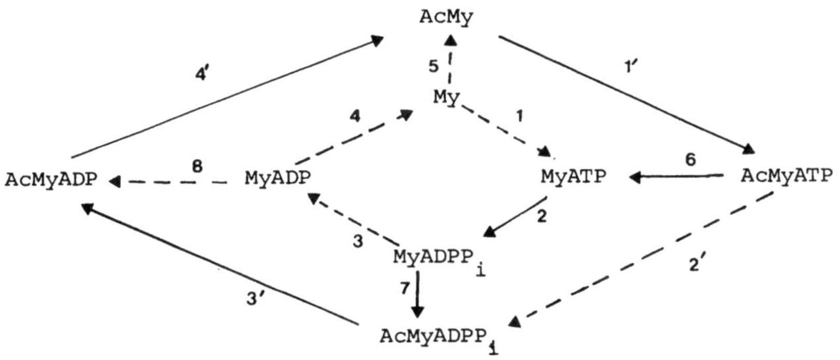

Myosin alone splits ATP (reactions 1, 2, 3, 4), but only slowly because reactions 3 and 4 are slow. This situation may be analagous to resting muscle in which interaction with actin is prevented. We therefore expect the myosin in resting muscle to exist largely as MyADP or $MyADPP_i$. When interaction of actin and myosin is allowed (active muscle, or addition of actin in vitro) an actomyosin complex is formed (reaction 7), and dissociation of the products, ADP and P_i, proceeds by reactions 3' and 4', thus bypassing the slow reactions 3 and 4. ATP binds to the actomyosin complex and then very rapidly dissociates it (reactions 1' and 6). ATP splitting occurs by reaction 2, not 2'. It is probably during the reactions of product release from actomyosin, that force is developed in active muscle. So we expect to find at least some of the myosin in active muscle in the forms AcMyADP and $AcMyADPP_i$. We thus expect a change in the state of the contractile proteins on transition from rest to activity in muscle.

EXPERIMENTS WITH LIVING MUSCLE

Figure 1 shows some examples of these experiments. The measurements of

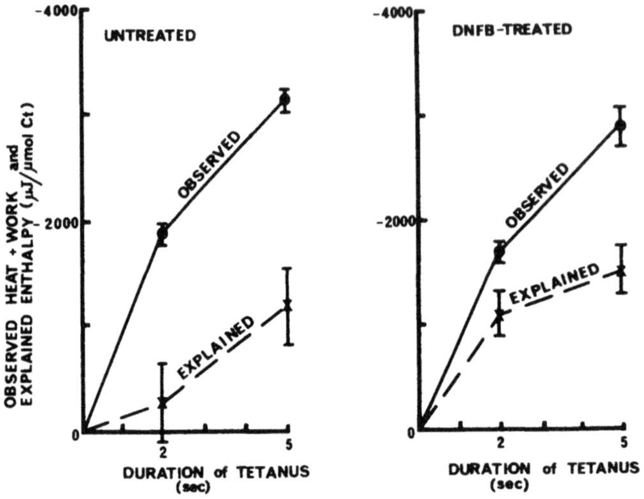

Figure 1.
From Curtin & Woledge, 1975. Measurements of energy produced during isometric tetanic contractions of frog (R.temporaria) sartorius muscles at $0^{o}C$. The full lines are the calorimetrically observed quantities. The broken lines show the amounts of energy production which can be explained by the changes in metabolite levels.

heat and work produced by the muscle (lines labelled 'observed') are made essentially by the techniques developed by A.V.Hill (3). To assess the heat produced by ATP splitting and the reactions which resynthesise it, a parallel series of experiments are done in which muscles are rapidly frozen during contraction and the metabolites extracted. The extracts are analysed for creatine, phosphocreatine, inorganic phosphate, ATP, ADP, AMP, inosine monophosphate, and a number of glycolytic intermediates. From the changes in concentration of these compounds the amounts of ATP split and the extent of various resynthetic reactions can be calculated. The heats of these reactions under in-vivo conditions are known (4, 5) and so the amount of energy they produce in the muscle can be calculated. The result is shown in the figure by the lines labelled 'explained'. There is no doubt that the explained energy is much less than the observed quantity. In other words a large part of the energy comes from reactions other than those considered in calculating the explained heat + work.

Some more findings about the production of unexplained energy by muscle will now be summarized briefly.

1. The phenomenon was first observed in frog muscle at $0^{o}C$ (6). It is more marked in Rana temporaria than in R. pipiens (7). It occurs also at 20° in frog muscle (8) and in rat muscle (9).

2. Unexplained energy continues to be produced during relaxation (10). Thus the process which produces this energy is not reversed in relaxation. We do not know when it is reversed, but studies of glycolytic and oxidative recovery have shown unexpectedly low amounts of rephosphorylation (11). This would be consistent with the reversal of the energy yielding process during recovery.

3. The unexplained energy occurs also in DNFB treated muscle (Fig. 1 and ref. 12) in which creatine kinase is inhibited. Thus it is not solely a result of the action of creatine kinase.

4. The amount of unexplained energy is greater when the muscle shortens rapidly than when contraction is isometric (no shortening) (13). This supports the suggestion that the contractile proteins are a source of unexplained energy since they would be expected to be directly influenced by change in the mechanical conditions during contraction.

5. The unexplained energy occurs also in contractions in which the muscles perform external work (14). Whether any of the work is derived from the unidentified process is not known.

It is now clear that a large part of the energy for muscle contractions under many conditions comes from reactions other than those usually considered as responsible for the supply of energy. These extra reactions could conceivably involve other metabolites for which we have not analysed. Although this possibility seems unlikely it is hard to eliminate it rigorously. It does seem more likely however that the unexplained energy comes from the proteins in the muscle. That this idea is quantitatively feasible can be seen from the following figures. The amount of unexplained energy after 5 sec of stimulation is 50 mJ/g of muscle. One gram of muscle contains about 0.2 µmol of myosin, that is 0.4 µmol of myosin heads. Thus if each myosin head were to undergo a change on activation of the muscle which produced 125 kJ/mol, the energy seen in the muscle could be explained. Energy changes of this magnitude are quite possible in protein molecules.

EXPERIMENTS ON MUSCLE PROTEINS IN VITRO

The work on intact muscle has suggested that hypothesis that the unexplained energy in muscle contraction comes from changes in the contractile proteins. Kinetic studies of these proteins suggest that changes in their state occur on activation. We have therefore begun to investigate the heat produced by these changes. We need to know the heat changes of all 12 of the reactions shown in the scheme for actomyosin ATPase above. Because of the interconnections of these reactions it will be necessary to determine only 7 of the heats of reaction, when the others can be obtained by subtraction using the heat of the overall process. So far we have information on only three of these reactions, which I shall now describe.

Interaction of Myosin and ADP

Since the equilibrium of reaction 4 favours the bound state this reaction is easily studied by adding ADP to myosin, when the reaction occurs in the reverse of its direction during ATP splitting. We have used a titration technique for these experiments which has the advantage that the equilibrium constant and stoichiometry of reaction can also be obtained in the same experiment.

An LKB batch microcalorimeter was used. It was modified for titration calorimetry by fitting a pair of motor driven micrometer syringes to the outside of the calorimeter drum (Figure 2). The syringes are connected by means of plastic capillary tubes to the usual cells. One cell contains myosin in buffer solution (about 60 nmol in 4 ml) the other cell contains buffer alone. After the calorimeter drum has been rotated several times, to equilibrate the surfaces of the cells with the solutions they contain, the experiment is made by injecting ADP solution (about 15 nmol in 4 µl) into both cells, and then rotating the drum to mix the reactants. 10 or 12 injections are made, interspersed with rotations of the drum without injection. A typical record of part of such an experiment is shown in Figure 3, and Table 1 shows the heats observed and their standard errors

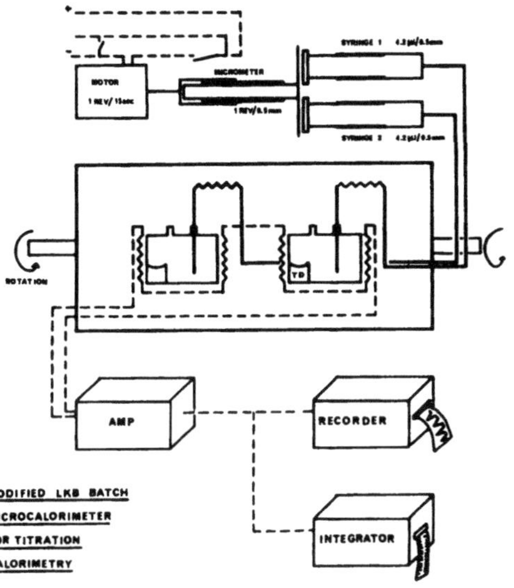

Figure 2. Diagram of apparatus used for titration calorimetry. The motor driven microsyringes are mounted on the outside of the calorimeter drum, within the thermostated air bath: Injections of titrant are made into both cells. The titrant (TD) in one cell; the other contains buffer solution only.

for two series of such experiments. The heat observed for one addition of ADP is about 1 mJ at the start of the titration and then diminishes as the myosin becomes saturated with ADP. In Figure 4 the total heat produced in a titration is plotted against the amount of ADP added.

If we assume that the myosin molecule has n independent and identical sites at which ADP is bound (with equilibrium constant K and heat of reaction ΔH) we can fit a theoretical curve to the calorimetric data, by means of an iterative least squares procedure using three adjustable parameters, n, K and ΔH. As seen in Figure 4 the calculated curves fit the data well. Table 2 shows the results of a number of series of experiments with myosin and with HMM. As expected the results are similar for both proteins, showing that removing the tail of the myosin molecule has little effect on ADP binding. The values of K are in reasonable agree-

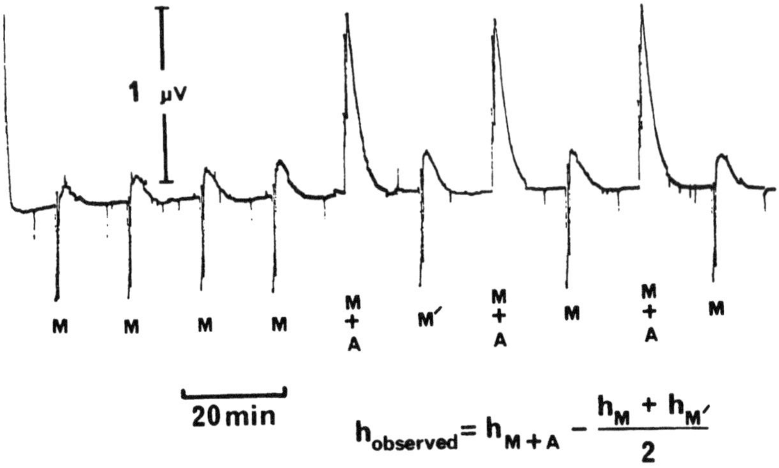

Figure 3. Part of the calorimeter record obtained during a titration of myosin with ADP. The calorimeter drum is rotated, to mix the cell contents, every 15 mins, marked by M. Some rotations are accompanied by an injection of ADP solution, marked by A. The heat attributable to reaction of ADP with myosin is calculated by the formula shown.

MYOSIN/ADP

ADP ADDITION	0°C [8] MEAN	0°C [8] S.E.	12°C [7] MEAN	12°C [7] S.E.
1	0.965	0.065	1.285	0.050
2	1.005	.050	1.237	.065
3	1.085	.029	1.097	.068
4	0.815	.065	1.127	.081
5	.695	.061	1.016	.068
6	.495	.078	0.880	.084
7	.245	.089	.622	.080
8	.125	.070	.309	.040
9	-0.065	.062	.275	.090
10	-0.035	.074	.172	.045
11	0.025	.060	.109	.042
12	—	—	.152	.050

Table 1. Heat (in mJ) observed at each addition of ADP during titration of myosin with ADP. Average results from 8 experiments at 0°C and 7 experiments at 12°C. 64 nmol of myosin in 4 ml of buffer was used in each experiment. At each ADP addition 17.4 nmol was added. Further details are given in Kodama & Woledge (18)

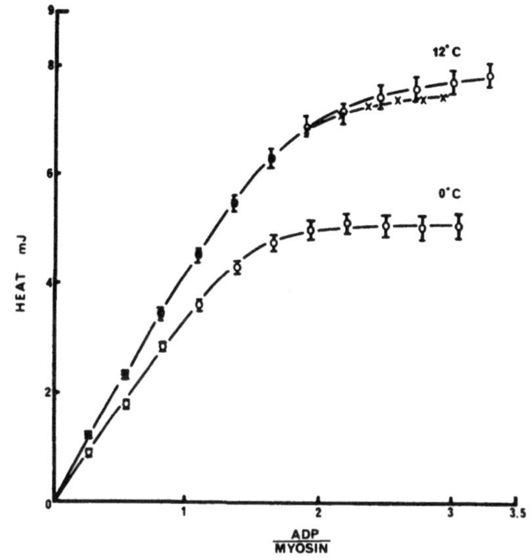

Figure 4. Results of calorimetric titration of myosin with ADP. The points, O, are mean experimental results, the bars show ± 1 S.E. The points x are the result of correcting the 12°C results for a side reaction, ADP splitting. No correction is needed at 0°C. The lines are theoretical curves calculated as described in the text. For further details see Kodama & Woledge (18).

PROTEINS	MYOSIN			HMM	
KCl CONC.	0.5M			0.5M	0.1M
TEMPERATURE	0°C	12°C	25°C	12°C	
ΔH [kJ mole^{-1}]	-57.1 ± 3.2	-73.1 ± 3.2	~ -70	-73.8 ± 1.3	-82.2 ± 1.9
log.K	6.42 ± 0.13	6.08 ± 0.13	5.5	5.80 ± 0.15	5.44 ± 0.19
n	1.49 ± 0.18	1.74 ± 0.07	—	1.75 ± 0.06	1.62 ± 0.09
n_{H^+}	0.18 ± 0.02	0.20 ± 0.02	—	—	—
$n_{H^+} \Delta H_b$ [kJ mole^{-1}]	-9.1 ± 1.0	-9.6 ± 1.0	—	—	—

Table 2.

ment with those obtained by other methods. However the values of n are always rather less than the expected value of 2. This could be the result of the presence of inactive protein, but an alternative possibility is that the second nucleotide molecule is bound more weakly than the first, so that true saturation is not achieved in our experiments. Further experiments are needed to examine this possibility.

In recent experiments with myosin agregated into filaments, and also preliminary observations with S_1, we have been able to observe very similar ΔH and K values for ADP binding. Thus under all the conditions we have investigated, ADP binding is strongly exothermic. By contrast Goodno and Swenson (15) have reproted that ADP binding to HMM produces only a small amount of heat. We cannot at present explain this discrepancy.

Interaction of Myosin with ATP

When a stoichiometric amount of ATP is added to myosin, reactions 1 and 2 occur rapidly so that $MyADPP_i$ is formed. The release of Pi from this complex (reaction 3) then occurs more slowly. As first shown by Yamada, Shimizu and Suga (16) these two phases can be distinguished calorimetrically. We have studied these reactions in a Calvet calorimeter continuously stirred by vertical oscillations of a perforated teflon disk. So far we have made only some preliminary experiments, one of these is shown in Figure 5 (open points). We find that there is no rapid phase of heat production but that all the heat is produced with an exponential time course of time constant about 80 sec, which is about the rate expected for reaction 3. If the starting material is MyADP instead of myosin (filled points in Figure 5) then there is a rapid initial heat absorption preceding the exponential heat production. This is due to the ATP 'chasing' the ADP from the myosin (reactions 4 and 1) which, as discussed above, is endothermic. It thus seems clear that reaction 3 is exothermic and that reactions 1 and 2 together are almost thermally neutral. Our first estimate of the heat of reaction 3 is -140 kJ/mol. This is in reasonable agreement with the results of Yamada et al.

The separation of the heats of reactions 1 and 2 is a difficult problem. Although reaction 2 is much slower than reaction 1 (at the usual ATP con-

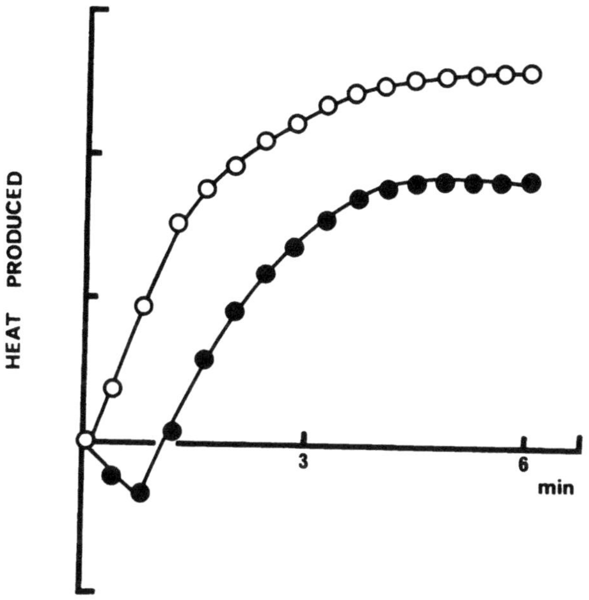

Figure 5. Result of calorimetric experiment on addition of ATP to myosin (O-O-O) or to myosin-ADP (●-●-●). Measurements were made with a Calvet calorimeter and have been corrected for heat loss and instrument lag by the methods of Hill (3).

centrations used) it is too rapid (400 s^{-1}) for the heats of the two reactions to be separatedly measured, at least with conventional calorimeters. Possibly the use of ATP analogues, such as ATPγS, for which reaction 2 is much slower may be helpful. Our expectation is that ATP binding (reaction 1) will turn out to be exothermic, like ADP binding, in which case the actual ATP splitting reaction (reaction 2) must be endothermic.

Interaction of Myosin with Actin

The equilibrium of reaction 5, the interaction of actin and myosin strongly favours the actomyosin complex. The heat of the reaction can be measured by mixing actin with HMM or S_1 in a batch calorimeter. We have used a variation of the titration technique used for studying ADP binding. Myosin (HMM or S_1) was placed in the titration syringe and the actin solution in the calorimeter cell. Because of the limited concentration of myosin that could be achieved, rather large volumes (100 to 200µl) had to be injected. These injections caused artefacts much larger than the expected heats of reaction. We therefore deferred mixing of the reactants until 10 mins after the injections. The type of record obtained is shown in Figure 6. Small heat absorptions were observed when the reactants were mixed; these were only a few tenths of a millijoule and near the limit of resolution of this technique. Nevertheless we were able to obtain reasonably consistent results with both HMM and S_1 which are shown in Figure 7. The results for HMM (two headed) and for S_1 (one headed) are very similar when plotted per mole of myosin heads. The similarity suggests that both heads of the HMM molecule attach to the actin filament, for which evidence was previously lacking.

We have attempted to analyse the results of the experiments on the assumption that the binding sites for myosin on the actin filament (n per actin monomer) are identical and independent. On this assumption we have calculated the best fit curves shown in Figure 7, using three adjustable parameters: K , ΔH, and n. Well fitting curves were obtained as shown in Figure 7. The best fit parameters were,

$$\text{for } S_1\text{: } n = 0.76 \quad \Delta H = +22 \text{ kJ/mol, } \log K = 6.6$$

$$\text{and for HMM: } n = 0.60, \; \Delta H = +38 \text{ kJ/mol, } \log K = 5.9$$

However these values for n and log K are much lower than expected on the basis of titration experiments using light scattering as an indicator of the extent of actomyosin formation (17). We therefore feel that the assumption of independent and identical sites is not correct. It seems likely that the attachment of the first myosin heads to the actin filament produces a larger heat absorption than do later attachments. The fact that the reaction is an endothermic one is however clear. In the case of HMM the average heat of reaction is close to +20 kJ/mol.

CONCLUSION

Our work on the calorimetric study of the ATPase system has only reached about a half way stage. We hope it will be possible in the next few years to find the heats of all 12 reactions involved in actomyosin ATPase. Only then will it be possible to assess the hypothesis that the origin of part

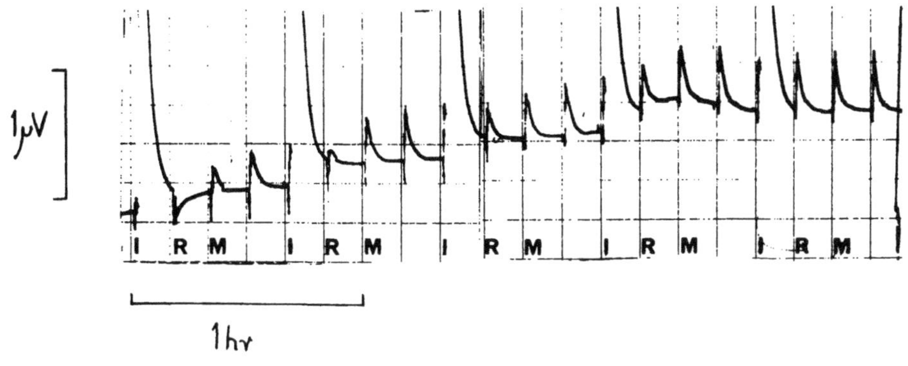

Figure 6. Part of the calorimeter record obtained during a titration of actin with HMM. The calorimeter is rotated every 10 min. Immediately after every 4th rotation (marked I) HMM solution is injected into one cell only but remains separate from the actin solution. This causes a large artefactual heat production. On the next rotation of the calorimeter (marked R) the actin and myosin solutions are mixed. The heat absorbed by the reaction of actin and myosin may be seen by comparison of the heat after this rotation with that after the subsequent rotation M. The amount of heat absorbed is clearly greatest in the first step of the titration and then decreases as the actin becomes saturated with myosin.

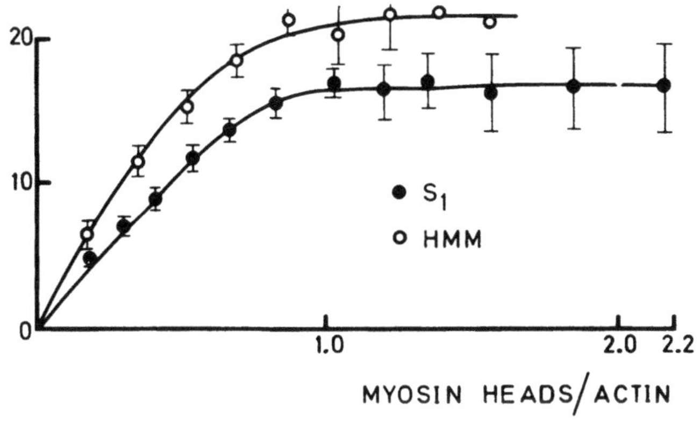

Figure 7. Results of calorimetric experiments on binding of myosin to actin. Points are mean results from 7 experiments with S_1, and 3 experiments with HMM. Bars show ± S.E. Lines are theoretical binding curves - see text.

of the energy appearing during muscle contraction comes from changes in the state of the contractile proteins. However at this stage it is at least apparent that some of these changes do produce enough heat to be potentially important sources of energy in muscle contraction.

REFERENCES

(1) Carlson, F.D., Wilkie, D.R.: Muscle Physiology. Prentice Hall Inc. (1974).

(2) Wilkie, D.R.: Muscle (Second Edition), Institute of Biology, Studies in Biology No. 11, Edward Arnold (1976).

(3) Hill, A.V.: Trails and Trials in Physiology, Edward Arnold (1965).

(4) Woledge, R.C.: Heat production and chemical change in muscle. Prog.Biophys.Molec.Biol., 22, 37-74 (1971)

(5) Woledge, R.C.: In vitro calorimetric studies relating to the interpretation of muscle heat experiments. Cold Spring Harbor Symposium 37, 629-634 (1972).

(6) Gilbert, C., Kretzschmar, K.M., Wilkie, D.R., Woledge, R.C.: Chemical change and energy output during muscular contractoin. J.Physiol. 218, 163-193 (1971).

(7) Homsher, E., Rall, J.A., Wallner, A., Ricchuiti, N.V.: Energy liberation and chemical change in frog skeletal muscle during single isometric tetanic contractions. J.Gen.Physiol. 65, 1-21 (1975).

(8) Canfield, P., Lebacq, J., Marechal, G.: Energy balance in frog sartorius muscle after an isometric tetanus at 20°C. J.Physiol. 232, 467-483 (1973).

(9) Gower, D., Kretzschmar, K.M.: Heat production and chemical change during isometric contraction of rat soleus muscle. J.Physiol. 258, 659-672 (1976).

(10) Curtin, N.A., Woledge, R.C.: Energetics of relaxation in frog muscle J.Physiol. 238, 437-446 (1974).

(11) Kushmerick, M.J., Paul, R.J.: Aerobic recovery metabolism following a single isometric tetanus in frog sartorius muscle at 0°C. J.Physiol. 254, 693-709 (1976).

(12) Curtin, N.A., Woledge, R.C.: Energy balance in DNFB-treated and untreated frog muscle. J.Physiol. 246, 737-752 (1975).

(13) Rall, J.A., Homsher, E., Wallner, A., Mommaerts, W.F.H.M. : A temporal dissociation of energy liberation and high energy phosphate splitting during shortening in frog skeletal muscles. J.Gen.Physiol. 68, 13-25 (1976).

(14) Curtin, N.A., Gilbert, C., Kretzschmar, K.M., Wilkie, D.R.: The effect of performance of work on the total energy output and metabolism during muscular contraction. J.Physiol. 238, 455-472 (1974).

(15) Goodno, C.G., Swenson, C.A.: Calorimetric studies on the binding of ADP to heavy meromyosin. J.Supramol.Struct. 3, 361-367 (1975).

(16) Yamada, T., Shimizu, H., Suga, H.: A kinetic study of the energy storing enzyme product complex in the hydrolysis of ATP by heavy meromyosin. Biochim.biophys.Acta. 305, 642-653 (1973).

(17) White, H., Taylor, E.: Biochemistry (in the Press) (1976).

(18) Kodama, T., Woledge, R.C.: Calorimetric studies of the interaction of myosin with ADP. J.Biol.Chem. (in the Press) (1976).

4.3 Calorimetric Investigations of Metabolic Regulation in Human Skin

A. Anders, G. Welge, B. Schaarschmidt,
I. Lamprecht, H. Schaefer

Introduction

In a series of experiments concerning the adaptation of the photodynamic effect to therapies of dermatologic diseases - especially the Psoriasis - and the application of dye lasers to this purpose (1) it was of some interest to get further information about the metabolism of healthy and diseased epidermis. Up to now experiments on the total metabolism of whole skin or different layers of it and on special steps in the glycolysis were only performed manometrically with the Warburg technique or enzymatically (2,3). Moreover, many of the manometric experiments require prior homogenisation of the samples, so that all regulative phenomena which take place due to the structural integrity of the cells were cancelled out. As far as we know no calorimetric investigations have ever been performed.

Van Scott has proposed a method by which the intact epidermis is stripped off from the cutis without effecting the horny layer (4). This procedure enables one to investigate in a system of complete cells the metabolic activity as well as the permeability of the membrane for enzymes, substrates, coenzymes and other molecules.

Psoriatic cells are well known for a high rate of replication and for a strongly increased metabolic turnover (5-8). The great advantage of a calorimeter is its unspecifity which enables one to detect even unexpected processes due to their enthalpy changes. Therefore, it is extremely suited to determine the differences between healthy and diseased tissues. In this paper we have to distinguish between two different kinds of experiments:

(a) The epidermis, streched in a buffer without any additional energy source, renders the endothermic metabolism.

(b) From the outside, an appropriate substrate is offered to the cells which is metabolized at a specific rate. This substrate may be glucose as an unspecific energy source or a substance that is utilized in only one specific metabolic step.

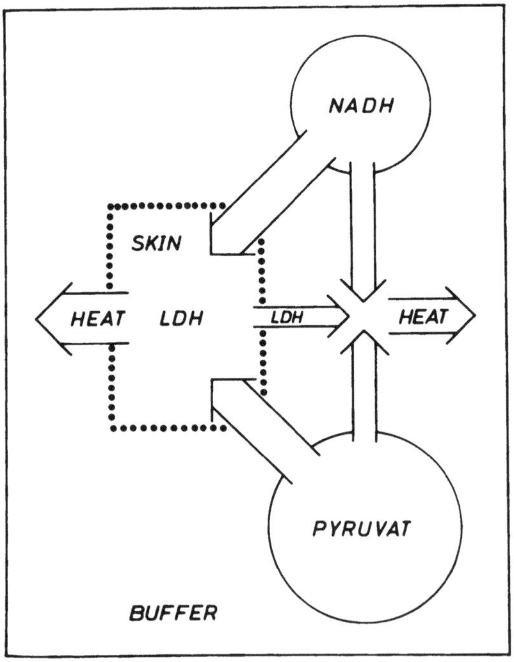

Fig.1

Flow of the substrate pyruvate and the coenzyme NADH into and outflow of unbound enzyme LDH from a skin specimen suspended in buffer and heat production by the hydrogenation of pyruvate within and outside the specimen.

In both cases, a necessary prerequisite is that the substrate can diffuse through the cell walls. At the same time a part of the free, unbound enzyme may leave the cell so that the reaction takes place inside and outside the tissue (Fig.1). When the specimen is later removed from the buffer and new substrate added an evaluation of the enzyme activity in the buffer is possible and a recalculation of the amount of intracellular enzyme. When the substrate is available in high concentrations the steady state metabolism is increased for a prolonged period. The value of the steady state may be influenced by some limiting factor within the cell, e.g. the permeability of the cell membrane for the substrate and coenzyme. Therefore, the rate of metabolism and the calculated enzyme activity do not necessarily correspond with the optimal activity (3).

On the other hand, if we are interested in the metabolic efficiency and the heat released per mole substrate, it would be more appropriate to apply a "ballistic method", i.e. a smaller amount of substrate which is consumed in an interval comparable with the time constant of the calorimeter. Since we are often confronted with first order reactions and the instrumental time constant is large compared with the rate constant a calibration with commercially available enzyme or a mathematical folding is necessary in order to compute the real turnover rate from the maximum heat flow and from the time interval between the start of the reaction and its maximum (9,10). Figure 2 demonstrates that only with LDH activity lower than appr. 7 U is there a linear relation between the activity and the heat production or the time τ (see below). With higher activity, both parameters are constant and would not allow the calculation of the real activity.

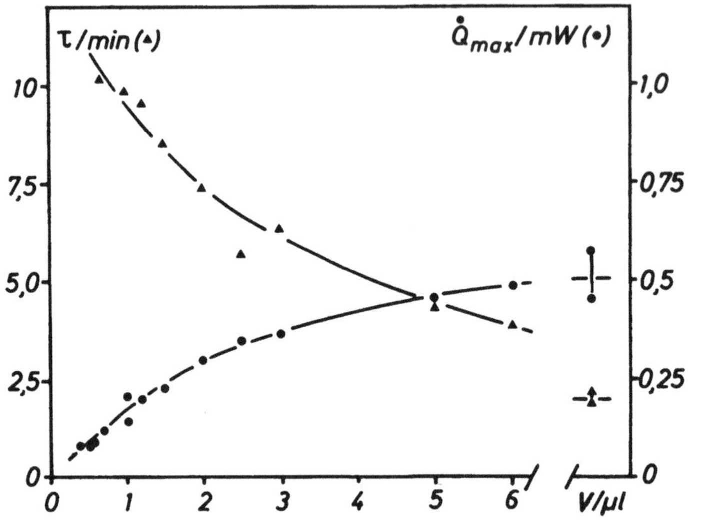

Fig. 2 Dependence of the maximum heat flow $\dot{Q}_{max}$ and the time τ from the addition of coenzyme to the maximum heat flow on the specific enzyme activity A in "ballistic experiments".

<u>Method and Material</u>

Fresh skin was obtained at breast cancer surgery or leg amputations and sometimes from corpses. The latter one is not as suited for metabolic experiments as the fresh skin samples. The epidermis is sheared off from the cutis following the method of van Scott (4). The samples were stored in a refrigerator at -18 C until used. For the tests, small pieces of 1 to 2 square centimeters (5-10/mg fresh weight) are placed in small frames with synthetic gauze windows so that the tissue stays stretched and offers a maximum surface to the surrounding buffer.

In former experiments the epidermis was compared to small liver slices, as these are more easily available. But cutting the slices injures the top layer of cells leading to a break down of regulative processes, whereas the inner cells do not obtain a sufficient supply of substrate. This fact led to the experiments with an isolated perfused rat liver which are reported by Baisch (these Proceedings).

Besides the glucose metabolism of cells in a bicarbonate buffer the main interest concerned the enzymatic activity of LDH - lactate dehydrogenase (EC 1.1.1.27) - which reduces pyruvate to lactate (11). The buffer contained the substrate pyruvate in excess while the coenzyme NADH was added in exact quantities and represented the limiting factor of the reaction.

The experiments were run in a Calvet microcalorimeter with 15 ml lucite vessels and a recorder sensitivity of 15 µW/cm. During the thermal equilibration the NADH solution was kept in a syringe just above the calorimetric vessel and at the same temperature. The syringe could be operated from outside the calorimeter. The temperature of the instrument was usually adjusted to 25 C, as at higher temperatures the lysis of the tissue is too rapid. An experiment on one specimen with 3 additions of NADH lasted a few hours. In this time there is no significant contamination by bacteria in the vessels.

To obtain the heat of reaction in vitro and to calibrate metabolic activities, commercially available LDH (Boehringer/Mannheim) in varying concentrations was used instead of epidermis.

Results

a) Pure Enzymatic Reactions

At high enzyme concentrations i.e. above 1 unit/ml we get a thermogram of a first order reaction, as may be shown by plotting the heat flow $\dot{Q}$ versus the amount of heat Q produced up to this moment (Fig.3). For a first order reaction this plot must be a straight

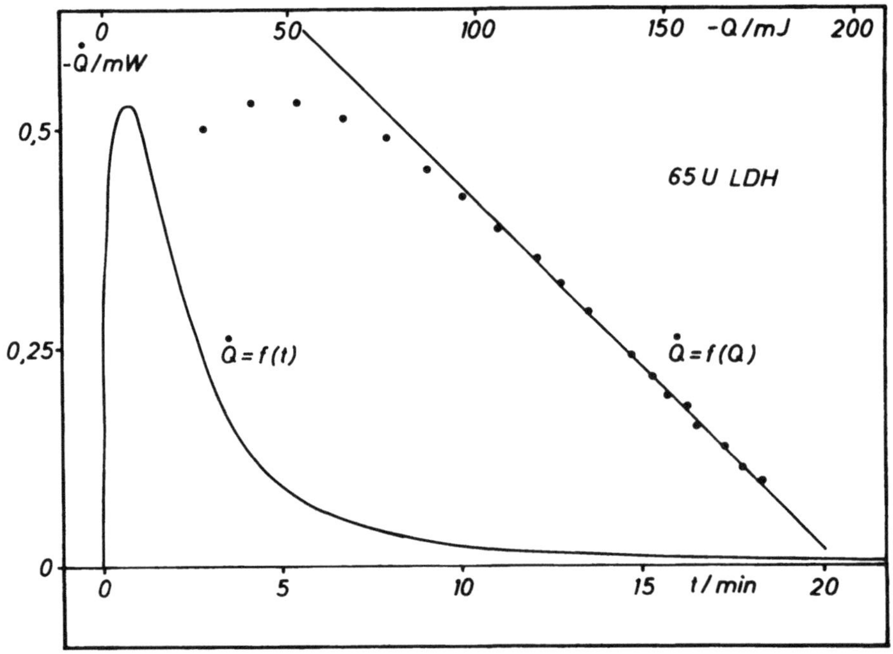

Fig. 3 Thermogram of the hydrogenation of pyruvate by means of 65 U commercially available LDH ($\dot{Q}$=f(t)) and its transformation to a linear first order plot ($\dot{Q}$=f(Q))

line. The deviation in the first part is due to the time constant of the calorimeter. The heat produced per mole NADH or per mole pyruvate hydrogenated to lactate is appr. 35 kJ, a figure corresponding to data from the literature.

b) Epidermis

At high concentrations of NADH and pyruvate, the maximum heat production in the steady state is a function of the enzyme activity in the tissue. Figure 4 shows two different ways to determine this figure: in the upper part the epidermis rests in a pyruvate buffer

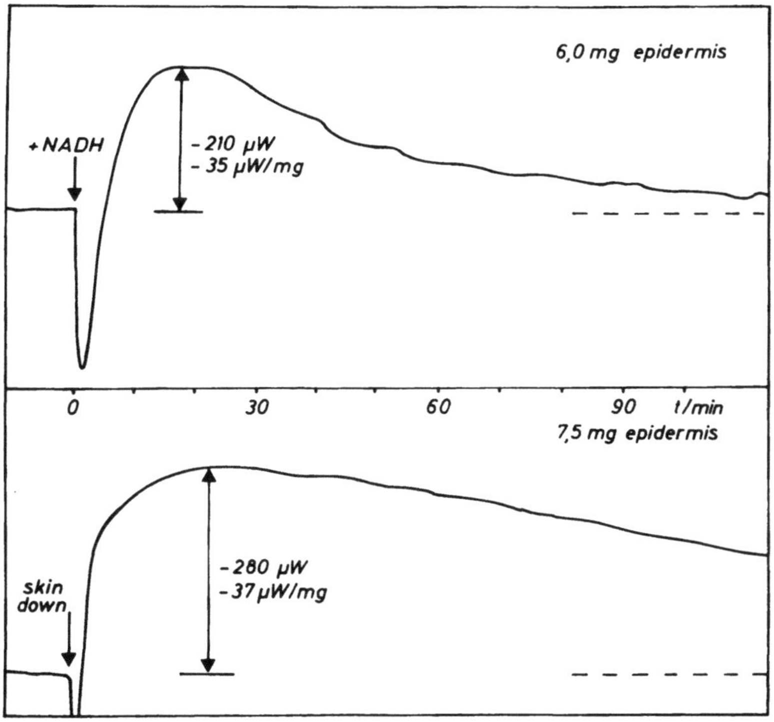

Fig. 4 Thermogram of the hydrogenation of pyruvate by two epidermis specimens. More details are given in the text.

without NADH. On addition of NADH (arrow) there is a short endothermic heat of dilution and then an exothermic heat flow due to the activity of cellular LDH. The area under the thermogram indicates a heat of hydrogenation of pyruvate which demonstrates that NADH is completely oxidized with the same efficiency as in vitro. The maximum heat flow corresponds to the maximum enzymatic turnover. In the lower part of the picture the buffer contains all necessary substances with the epidermis suspended above it. When it is submersed an equivalent heat flow is observed which stays constant over a prolonged period as the NADH concentration is four times that in the first experiment. This higher concentration of NADH could be used since there is no problem with a heat of dilution. From both experiments a maximum heat flow of appr. 35 µW/mg wet weight of epidermis is calculated. Under the assumption that the enzyme activity and not a transport factor is the limiting step of this reaction one gets a minimum enzyme activity of appr. 60 mU/mg, which is in good agreement with data for other organs given in the literature. Similar results are obtained if glucose is added to the system or if the epidermis is submersed in a glucose buffer.

In experiments with consecutive addition of NADH to epidermis in buffer, the heat output in each run is nearly constant and equals that of the enzymatic reaction in vitro. Before the last addition of NADH the epidermis may be pulled out, so that only the LDH which diffused from the tissue into the buffer is able to hydrogenate pyruvate. Since the endogeneous metabolism of the skin probe still continues, it is adviseable to remove the epidermis from the calorimetric vessel and then check the enzyme activity in the buffer. The remaining activity proofed that a large amount of LDH leaves the cells. Schaefer and coworkers (3) have found photometrically that 70 per cent of the enzyme activity leaks which seems to agree with the present data.

Conclusion

These preliminary experiments show that calorimetric investigations are suited for the evaluation of epidermal metabolism. Further experiments will be aimed at other steps in glycolysis, in the citric acid cycle and in the respiration chain. Their results will be published elsewhere (12). Moreover, the influence of photodynamically active dyes - such as thiopyronine and 8-methoxypsoralen - on the metabolism of healthy tissue and of diseased skin or on special metabolic steps is of great interest for the dermatologic therapy. We thus hope to find a further approach to the photodynamic therapy of dermatosis and a deeper understanding of this effect.

Acknowledgement: The calorimeter was placed at our disposal by the Deutsche Forschungsgemeinschaft/Bonn-Bad Godesberg

References

1 Anders,A., Lamprecht,I., Schaefer,H., Zacharias,H.:
The use of dye lasers for spectroscopic investigations and photodynamic therapy of human skin
Arch.Derm.Res. 255, 211-214 (1976).

2 Laerum,O.D., Bjerknes,R.:
Seasonal and individual metabolic variations of hairless mouse skin
J.invest.Derm. 58, 284-290 (1972).

3 Schaefer,H., Zesch,A., Stüttgen,G.:
Methodik zur fotometrischen Messung von Enzymaktivitäten in intakten menschlichen Epidermiszellen
Arch.klin.exp.Derm. 239, 347-354 (1971).

4 Scott, J.J.van:
Mechanical separation of the epidermis from the corium
J.invest.Derm. 18, 377-379 (1952).

5 Gans,O., Glasenapp,I. von:
Contributi al problema della psoriasi respitazione cutanea e glicolise al livello delle chiazze de psoriasi
Dermatologia (Napoli) 2, 1-2 (1951)

6 Herdenstam,C.G.:
On the in vitro metabolism of labeled glucose in normal and psoriatic skin slices
Acta derm.-venerol. (Stockh.) 42, Suppl.47, 1-49 (1962)

7 Comaish,I.S.:
Epidermal aldolase levels in psoriasis and normal skin
Brit.J.Derm.Syph. 75, 337-343 (1963)

8 Halprin,K.M., Ohkawara,A.:
Lactate production and lactate dehydrogenase in the human epidermis
J.invest.Derm. 47, 222-226 (1966)

9 Calvet,E., Camia,F.:
Sur l'obtention des courbes de thermogenèse à partir des courbes enregistrées au microcalorimètre de E.Calvet
J.Chim.Phys. 55, 818-826 (1958)

10 Zahra,C., Lagarde,L., Romanetti,R.:
Etudes thermocinétiques de réactions lentes par microcalorimétrie à conduction
Thermochim.Acta 6, 145-163 (1973)

11 Everse,J., Kaplan,N.O.:
Lactate dehydrogenases: structure and function.
in: Advances in Enzymology (Ed.A.Meister), Vol. 37, 61-133 (1973)

12 A.Anders, B.Schaarschmidt, H.Schaefer, I.Lamprecht
Microcalorimetric Investigations of the metabolism of isolated human epidermis
J.invest.Dermatol., submitted

4.4 Organperfusion in a Calvet Microcalorimeter: Adaptation of a Process Computer to an Experimental Arrangement Especially Designed for this Purpose

F. Baisch

Summary

A system analysis, including the temperature regulation, the perfusion medium, and the systematic error caused by the friction in the flowline is given. The process computer controls the experiment, correlates different process variables, e.g. flow speed and O_2-concentration of the perfusion medium, documents in EDP-usual data size (perforated tapes) and records with dot printer and telegraph.

The purpose of the apparatus is to establish reproducible experimental conditions for pharmacological and kinetic research on the metabolism of perfused rat livers.

Introduction

The aim of this paper is to give preliminary results on experiments concerning the metabolism of an isolated perfused rat liver. Although a series of experiments has been performed, it seemed to us more interesting to give a more detailed description of the instrumental setup and the difficulties which must be overcome in this research. Once the method is established, it is quite easy to test the influence of a variety of drugs, pharmaca, antibiotics and metabolites on the metabolism of the liver. Moreover, the results obtained with the intact liver and liver slices may be compared, thus providing a calibration for the data from isolated and sliced tissues [1,2].

Calorimetric research on tissues and on whole organs is intermediate between that on microbes and on whole organisms and involves special problems. If slices of tissues are cut off or single cells are separated by trypsinogen, they acquire new artificial surfaces and lose part of their regulation mechanisms. Many of the membranebound enzyms are distributed to the surrounding buffer system, the diffusion of internal cell enzymes is far higher and many cells are more like a homogenisate than intact cells. Under these conditions measurements of enzyme activities have little significance.

As liver probes had been chosen as comparative systems for calorimetric skin measurements [3] we decided to investigate the metabolism of an intact isolated and perfused rat liver. As far as we know, this is the first time that such an experiment has been carried out in a calorimeter, and as there were technical difficulties because of aeration of the liver and heat leakages by the perfusion liquid, we want to emphasize the technical part of these experiments.

Methods and techniques

All the experiments were done on a CALVET microcalorimeter, type 1203 N° 04 114 (SETARAM/Lyon) with two vessels of 100 ml each and a sensitivity of appr. 55µV/mW, and all equipments was designed for this instrument.

In the first picture, a liver is symbolically indicated by the marked square. In the living organism, it is supplied as follows:

The inflow (1) carries arterial blood that comes from the aorta via several branchings. This constitutes a high-pressure supply. The supply (2) represents the blood from the intestinal vessel collected in the portal vein. Since this blood has already flowed through a capillary network, the pressure is much lower. The slightest pressure is found in the outflow (3). It flows through a short vein into the right vestibule. In case of disease (cirrhosis, fatty liver due to alcohol) and, unfortunately, also in our experiments, part of the liquid flowing through the liver drains off through the parenchyma via liver surface. Thus, the pressure in the abdominal space, or as in our experiments, the pressure in the reaction vessel, cannot be neglected. But, otherwise, the classical perfusion technique is applied [4,5,6] . The liver, however, is totally excised for this purpose and placed on a support in the calorimetric vessel. For a short period of time (about 8 hours), the high-pressure supply can be removed without causing severe liver damage, especially when the portal vein blood is oxygenated. Even in vivo, the arterial inflow contains only 15% of the necessary oxygen. This can be compensated in vitro by a high oxygenation of the perfusion solution.

Consequently, for a simplified model the following conditions are necessary in order to guarantee hepatic functioning in the absence of the organism.

1. Physiological pH and oncotic pressure in the perfusion solution are maintained by means of organic buffers, salts and protein.
2. Sufficient oxygen is supplied and
3. Excess CO_2 is eliminated by an oxygenizer perfused with carbogenic gas (95% O_2, 5% CO_2).
4. Elimination of metabolic catabolites which are normally excreted

by the kidneys, and

5. Nutritive substances and pharmaceuticals to be tested can be exchanged by way of dialysis membranes.
6. Physiological perfusion pressure and
7. Relatively constant temperature and maintained by the instrumental setup described below.

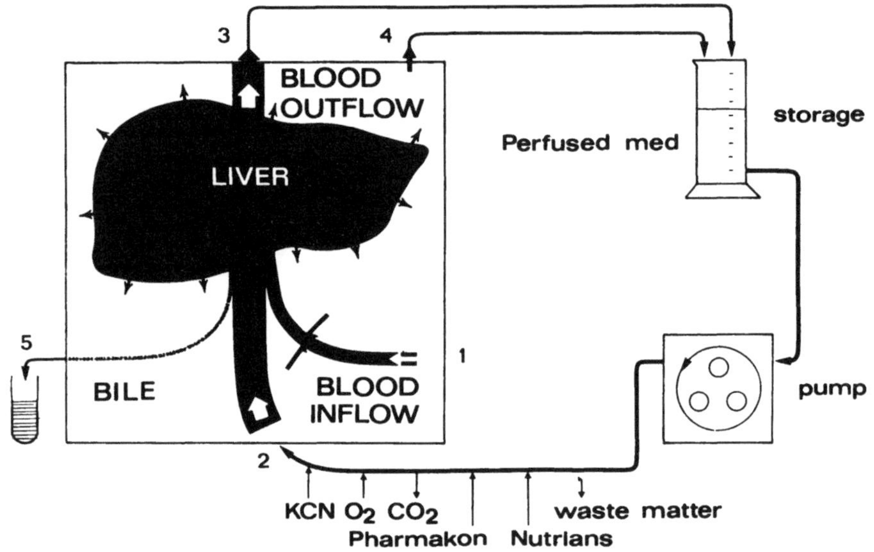

Fig.1 Perfusion flow system.
The liver is connected as follows:

Inflow : (1) Arterial supply. It is ligatured for the isolated liver.
(2) Portal vein

Outflow: (3) (Lower large) vena cava inferior
(4) Vessel outflow
(5) Bile outflow

In the inflow line there may be an exchange of gases, metabolites and waste matter via dialysis membranes. Drugs are added to this line by means of syringes.

We mention temperature last,since, in the classical perfusion technique, it is subject to the greatest deviation from the value in vivo. The liver is undercooled in order to suppress its activity, the medical term for this being "hibernation." 20°C and 30°C are common temperatures for the perfusion solution. Simplified according to van't Hoff, this means a quarter or half of the usual metabolic capacity. When these conditions are met, there is no necessity for erythrocytes as oxygen carriers.

For calorimetric investigations, temperature cannot be neglected. With ideal experimental conditions, the perfusion medium must flow isothermally into the reaction vessel of the calorimeter. This can be approximately achieved only by a cascade regulation. In order to balance out the last thousandths of a degree, the technical expense must be driven very high, especially since the medium is flowing, and for the rat liver a control flow of 10-20 cm^3/min. must be dealt with. For velocities up to 20 cm/s, which we observed in the afferent capillaries, the controlled system must be 20 cm long, in order to compensate within one second for a possible difference in temperature.

In the classical perfusion devices, the liver is perfused by means of hydrostatic pressure. Since there is a long way into the calorimeter, this difference in pressure cannot be transferred into the calorimetric vessel by communicating tubes.

Experiments in which the liver is perfused with a constant flow by means of peristaltic pump are unphysiological. In many cases, disruption of the tissue is the consequence, and the liver will no longer be able to regulate its perfusion independently. In order to avoid these difficulties, it is necessary to determine the pressure value between liver inflow and outflow, and to regulate the output of the peristaltic pump accordingly.

At present, the controlled system is positioned in such a way that the flow decreases to a tenth when the pressure is doubled. Further experiments are needed to indicate whether this linearly controlled system can be maintained. The pressure receiver for the control of the vessel pressure is shown in Fig.2. Moreover, the last cascade of temperature control is recognizable. The difference in temperature between the inflowing medium and the block of the Calvet calorimeter is measured.

The bridge tension arising in case of a deviation is reinforced and electric power is delivered accordingly to a resistance heater. At the moment, we perform experiments with a glass spiral which is wound with resistance wire. The perfusion medium flows through the spiral, penetrates the reaction vessel via a passive heat exchanger and perfuses the liver. Then it is led within spirals along the inside of the reaction cylinder and passes into a sort of "backflow heater" for the inflowing medium above the aforementioned heating spiral.

This route is a kind of inverse feedback. When the phase relationship between temperature difference and heater current is right, this system is effective. However, phase shifting is flow-dependent, and liquid flow has another effect in itself as can be seen from Fig.3. It shows the inner pressure P along the distance S which must be covered by the perfusion medium via the communicating tubes of the device: from the collecting vessel through a pump, dialysis membranes, different heaters into the calorimetric vessel and through the liver, and then back by the same route. From this point, it drops back into the vessel (V.4.2.and V.4.1.→ 0.5).

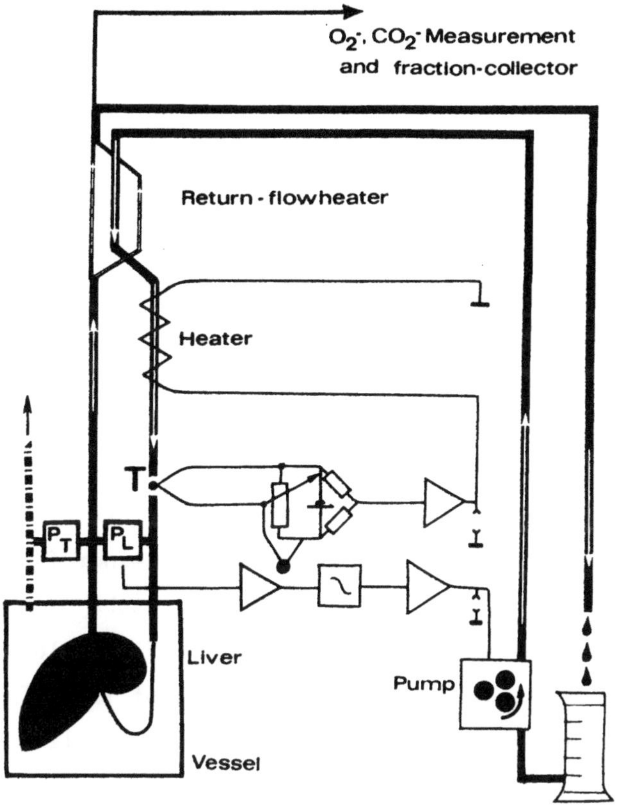

Fig.2 Control-loop for pressure and temperature

The pressure transducer P_L monitors the difference between inflow and outflow, P_T between outflow and reaction vessel. An electronic filter suppresses unwanted oscillations of the pump control. The temperature of the inflowing medium is measured a few centimeters above the entrance into the reaction vessel and compared with that of the heat sink of the Calvet calorimeter. The probe T is placed in a bulge of the glass tube leading to the liver. Thus it is situated in the flow without disturbing it. The heater is a glass helix wrapped with a resistance wire.

But this is only the case if the flow velocity tends towards zero and if there is no pressure loss due to flow friction. Actually, the pump must produce a high pressure in order to cause a pressure difference in the liver, despite the previously mentioned devices the medium flows through. Within the liver, the flow forms branches. Part of the liquid flows off over the liver surface, and in all our experiments, this constituted a non-negligible percentage of the overall flow. In such a total operation, small capsular lesions due to the liver preparation are unavoidable. Besides, the organ would tampon itself if the vessel pressure were not adjusted. Thus, the conduction at the discharge end must be lowered so far that the pressure differences in the liver are maintained when flow velocities of about 20 cm/min are present. With laminar flows (Reynold number smaller than 2320) and constant tube diameter, the pressure gradient depends on the square of the flow velocity (and on geodetical conditions which are independent of flow).This pressure loss is converted into heat, which is added to all our metabolic heat measurements as a systematic error. Our studies showed that the flow heat has the same order of magnitude as the heat production by metabolism.

This means that the error produced by friction and by the change in friction due to varying flows has to be eliminated. For this purpose, some kind of "line characteristic curve" for the liver must be ascertained. I borrowed this term from the technicians in line systems for heating and air conditioning, and it refers to nothing else but to the curve path: necessary discharge pressure as a function of the amount flow. As the inner surface of our reaction vessel is two-thirds covered by a double coil 1.30 m in length, the heat of friction developed therein must also be considered.

However, this method of leading the liquid which drains off from the liver along the inside of the reaction cylinder is needed to insure that all the heat produced in the organ is measured in the calorimeter [7].

The additional heat of flow in this coil can be calculated. Consequently, the heat due to flow friction is composed of the sum of separate components: firstly, the frictional heat along the two liver paths and secondly, the heat developing along the cylinder coils.

Only if this value is deduced point by point from the total heat flow, can a curve path be developed with which the actual metabolic capacity can be assessed. The second coordinate arrow(p_{at}) indicated in the y-direction on Fig.3 is to show that the perfusion circulation as a whole can be set under a higher than atmospheric pressure. This possibility is provided because with higher pressure, more oxygen is dissolved in the perfusion solution and therefore available for the liver. With an overpressure of 1.8 atmospheres, enough oxygen is dissolved so that the erythrocytes of a mammalian organism no longer participate in the gas exchange. In our case, this means that we can abandon the artificial hibernation technique and achieve higher metabolic rates at higher temperatures [8,9]. With sufficient oxygen supply, the perfusion flow even decreases, so that we are able to distinguish clearly the metabolic signal from the disturbing frictional heat.

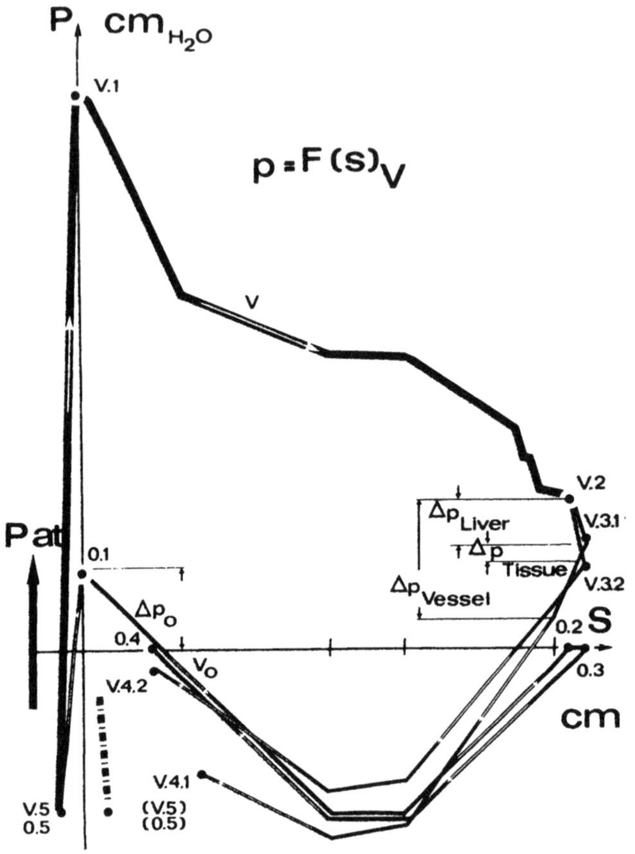

Fig.3

The abscissa represents the distance the medium flows, the ordinate the pressure on the inner surface of the tube. From point 0.5 or V.5 the pressure is increased to V.1 by means of a peristaltic pump at a flow rate V. From V.1 the pressure on the inner surface or the tubes decreases to the point V.2. From the liver exit V.3.1 and the reaction vessel V.3.2 the medium flows back to points V.4.1 and V.4.2. From here it drops back into the storage container.

However, we intend to minimize the amount of liquid in the perfusion cycle, so that the liver may act with a volume similar to that of the blood and extravasal liquid of the donor animal (appr. 10 to 20 ml).

Summarizing the requirements necessary for the perfusion device, the following activities must be presupposed in order to obtain reproducible values.

I. Prior to the test:

1. Preparation of the rat liver.
2. Control of the perfusion solution.
3. Control of the oxygenizer function.
4. Care of the artificial kidney.

II. During the test:

5. Control of the perfusion pressure.
6. Checking of the automatic control systems of temperature for all thermostats and the corresponding water-bearing ducts.
7. Well-timed switching on of the recording.
8. Sampling and control of the CO_2 and O_2 recording.
9. After a suitable time, starting of a slow increase of the overall pressure of the perfusion circulation with all difficulties of leakage control.
10. Addition of substances to be tested.
11. Recording of the individual terms.
12. Poisoning of the liver at the end of the test to get the experimental zero line. As few of the perfusion line as possible should come into contact with the poison.
13. Conversion of the extremely long registered curves in order to assess wether the experiment is usable at all.

14. If this is the case, the new curve path must be determined point by point.

Since this would require too much time or too many coworkers, we plan to assign all the above points except liver preparation to a process calculator. All data from measurement and limiting values are read in cyclically and, accordingly, the previously mentioned devices are connected or disconnected. The dot printer shows the instantly corrected curve path - and that seems to be most important for us.

All primary test data are present, corresponding to the computer data. This means that a test procedure may be simulated again at any time, and that the data may be combined according to other points of view. The procedure is fixed for one set of experiments. With only one parameter being changed, the problems, most of which can be solved only empirically, may be more easily eliminated, thus enabling an optimal test setup and procedure.

Literature

[1] De Venanzi, F., Pena, F., Jimenez, V.O., De Alvarado, H.: Effect of glucagon, epinephrine, cyclic 3', 5'-AMP, N^6-2'-o-dibutyryl cyclic 3', 5'-AMP and insulin upon the phosphate exchange of the isolated perfused fed rat liver. Endocrinology 95, 741-746 (1974).

[2] Woodside, K.H., Ward, W.F., Mortimore, G.E.: Effects of glucagon on general protein degradation and synthesis in perfused rat liver. J. Biol. Chem. 249, 5458-5462 (1974).

[3] Anders, A. et al.: Calorimetric investigations of metabolic regulation in human skin. This book.

[4] Miller, L.L., Bly, C.G., Watson, M.L., Bale, N.E.: The dominant role of the liver in plasma protein synthesis. J. exp. Med. 94, 431-453 (1951).

[5] Staib, W., Scholz, R.: Stoffwechsel der isoliert perfundierten Leber. Springer Verlag, Berlin (1968).

[6] Höllerer, M., Breuer, H.: Eine optimale Methode zur hämoglobinfreien Perfusion der isolierten Rattenleber. Z. Klin. Chem. 12, 398-402 (1974).

[7] Klinger, H.G.: Heat transfer in perfused biological tissue. Bull. Math. Biol. V3C, 403-415 (1974).

[8] Harms, H., Kirsch, U.: Hyperbare Oxygenation. In: Praxis der Intensivbehandlung, Ed. P. Lavin, 3. Aufl. Kap. 21, Thieme Verlag, Stuttgart (1975).

[9] Reisewitz, H.: Nierenfunktion während des totalen Blutersatzes mit einer Fluor-carbon-Emulsion bei den Ratten. Dissertation Fachbereich Medizin der Universität Hamburg, Hamburg (1974).

5. Medical Calorimetry

5.1 Applications of Microcalorimetry in the Medical Field

I. Wadsö

INTRODUCTION

Practically all processes are accompanied by heat effects. Calorimetry thus forms a general and quantitative analytical tool for simple chemical reactions as well as for an assembly of processes in a cellular system. The fact that calorimetric methods are so general makes them also very unspecific. This is a serious limitation for their practical use in many types of analytical problems. However, in biochemistry and biology the inherent specificity of the reaction systems themselves often allows the use of an unspecific analytical method. One may also note that, in particular for very complex systems, it is sometimes advantageous to use an unspecific method rather than a very specific analytical method since it is then more likely that unknown phenomena will be discovered. For living systems we may further note that life processes, when observed calorimetrically, are self-recording and that it is not necessary to disturb the biological system, e.g. by addition of reagents. Another factor of importance to investigations on bio-systems is that calorimetric methods, in contrast to spectrophotometric methods, do not require optically clear solutions and can be used with tissues or suspensions.

When an unspecific analytical method is used on a complex process it is necessary to make sure that the property actually measured is adequately related to that property or phenomenon one has decided to study. As heat effects are evolved at all kinds of processes calorimetry is particularly vulnerable to systematic errors or misinterpretations (1). This is not the least the case in experiments on complex systems which have not been very thoroughly evaluated, e.g. most cellular systems.

In modern clinical work chemical analysis plays a key role in the diagnosis of most diseases and for the control of results of medical treatment:

determinations of concentrations of metabolites, of enzymes and other proteins and of various cofactors. A large variety of instrumental procedures are utilized: spectrophotometry, electrophoresis, radioactive counting techniques etc. but calorimetry is as yet of no practical importance in this field. The instrumental methods in current use in clinical chemical analysis have reached a high level of technical development and one can conclude that this is a very competitive field. However, there are comparatively few instrumental methods which are useful for the characterization of cellular materials. It is therefore judged that calorimetry will prove to be more important for practical analytical work on this level than on the chemical level. Recent microcalorimetric studies on bacteria, blood cells and tissues certainly point in the direction of interesting new analytical determinations.

Many research groups rely today upon calorimeters as key instruments for bio-thermodynamic investigations. The impact on the applied areas by such studies cannot easily be assessed, but their importance for the medical field should not be neglected. For instance, studies of the thermodynamics of interaction between biopolymers and low molecular compounds (including drugs), of reactions involving immunochemical systems, of unfolding reactions etc., all contribute to a deeper understanding of processes which are of a fundamental importance for medicine. However, the present review will be concerned mainly with the use of microcalorimetry as a general analytical tool. Microcalorimetry and its use in biochemistry and biology as a thermodynamic tool and as an analytical tool have recently been discussed elsewhere (1). Microcalorimetric instrumentation was discussed in some detail (1), cf. also (2), and will therefore not be dealt with in this paper.

WORK ON THE BIOCHEMICAL LEVEL

Enzyme Assays

Benzinger and Kitzinger seem to have been the first to demonstrate the usefulness of calorimetry for determination of enzyme activity (3). These workers used their batch calorimetric method to follow the purification of a glutaminase preparation which was contaminated by carboxylase. Using glutamic acid as a substrate the decline in the undesired enzyme activity could be followed. Another early study was the one by Brown et al. on ATPase activity in erythrocyte ghosts (4).

Using a flow microcalorimetric technique Monk and Wadsö (5) reported a method for the assay of a number of clinically important enzymes. Solutions of glucose oxidase, choline esterase, alkaline phosphatase, lactic acid dehydrogenase were investigated as well as a tissue homogenate with ATPase activity. The enzyme preparation was mixed with an excess of substrate to give a zero order reaction system which was pumped through the flow through cell of the calorimeter. The displacement of the baseline was directly proportional to the heat effect and thus to the enzyme activity. The possibilities for "chemical amplification" were discussed, cf. (1), and in particular the use of a suitable buffer system was demonstrated. A typical example is provided by hydrolysis of an ester, e.g. acetyl cholin,

(1) $(ROAc + H_2O \rightarrow ROH + Ac^- + H^+)aq$

The enthalpy change of reaction (1) is close to zero, $\Delta H = 1.2$ kJ mol^{-1}. However, if the proton released is taken up by an amine buffer with a highly exothermic enthalpy of protonation (e.g. tris with $\Delta H_{prot} = -47.5$ kJ mol^{-1}, the reaction will become strongly exothermic. If the hydrolysis process is carried out in a phosphate buffer ($\Delta H_{prot} = -4.7$ kJ mol^{-1}) the measured reaction will be only weakly exothermic.

Several other microcalorimetric studies of enzyme activity measurements have since been reported e.g. (6)(7)(8)(9); see also the recent batch calorimetric studies by Rehak et al. (10) involving enzymes in solution as well as immobilized enzymes.

Determination of Substrate Concentration

It has been shown that nearly the same flow calorimetric technique as was used for enzyme assays can be adopted for substrate determination, the difference being that the substrate concentration should be rate limiting (5)(7). In principle, however, it is easier to determine substrate quantities in a batch or a stopped flow experiment. The heat quantity evolved in the experiment, after small corrections have been applied, is directly proportional to the quantity of substrate in the sample and the result is thus independent of the activity of the enzyme prepration used. Goldberg et al (11) have applied a batch calorimetric procedure for the determination of glucose in serum using the hexokinase reaction. These workers have also looked into the thermochemistry of this process in some detail. Glothlin and Jordan (12)(13) employed the same reaction for glucose determination in samples of serum, plasma or whole blood. These workers used an injection procedure with a simple Dewar vessel calorimeter.

The precision obtained in these calorimetric substrate determinations appears to be similar to that found with conventional techniques. Johansson et al. (14) demonstrated the use of immobilized enzymes in flow microcalorimetric determination of substrate quantities. An LKB flow calorimeter was equipped with a modified flow cell holding a column charged with about 0.3 ml of immobilized enzyme. When a solution containing a substrate was pumped through the system the reaction took place and a peak was recorded. The area under the peak is directly proportional to the heat quantity evolved and thus to the amount of substrate. More recently simple thermistor operated flow instruments for substrate determinations using immobilized enzymes have been designed by Mosbach and coworkers (15,16,17) by Canning and Carr (18) and by Schmidt et al. (19). The method has successfully been used for determination of e.g. glucose, lactase, uric acid, urea, penicillin and cholesterol. The attainable precision varies somewhat with the system investigated but it appears to approach that obtained with a more elaborate calorimetric technique. It is judged that this technique will soon be established in routine substrate analysis.

Plasma Coagulation

Plasma coagulation is the result of a very complex series of processes for which thermal observation methods seem to be suitable. Bostic and Carr (20) measured clotting time with a thermometric method and their thermograms also indicated the presence of different reaction steps. Watt et al. (21) used a more developed calorimetric technique and determined clotting time with a good precision (± SD was usually about 3 s). They also found characteristic heat effect patterns before and after the coagulation period. It seems as calorimetric investigations on the plasma coagulation system can yield new information and that the method can develop into a clinically useful technique.

WORK ON CELLULAR SYSTEMS

Microorganisms

Calorimetric methods have been shown to provide realistic alternatives to existing methods for the detection and enumeration of microorganisms as well as for their identification. Calorimetry can also provide continuous information about aerobic as well as anaerobic microbial growth processes which may be utilized e.g. in rapid antibiotic sensitivity tests. It seems very likely that microcalorimetric methods will be developed into clinically useful methods as well as in corresponding research areas.

Microbial activity. The use of microcalorimetry for detection of microorganisms and for their enumeration is presently being explored in several areas (1). A clinical application has been reported by Beezer et al. (22) who investigated the potential use of flow calorimetry for the detection of bacteriuria. Samples to be tested contained originally 10^3 - 10^5 bacteria$\cdot$ml^{-1}. They were mixed with a nutrient broth and were allowed to grow overnight under standardized conditions before being presented to the calorimeter. Results showed that it is possible to discriminate between different levels of initial bacterial concentrations (10^3, 10^4 or 10^5 bacteria$\cdot$ml^{-1}). It was concluded that the technique would lend itself for automatization in addition to the possiblity of extending the calorimetric experiment to the identification of the bacteria and subsequent antibiotic sensitivity tests (cf. below). It seems, however, as if the rapid chemical tests now used for the detection of bacteriuria will be difficult to compete with; probably calorimetry is more promising in connection with bacterial infection in blood, in particular when coupled to identification of the organisms from their growth patterns (23).

Regardless of the detection principle used it is necessary in clinical work to grow the microorganisms for some time before they can be enumerated. In studies of the optimization of growth media composition calorimetry appears to be an ideal method.

Identification of organisms. Calorimetrically recorded growth patterns for bacteria can be used as a method for their identification as demonstrated by Russel and his associates (24)(25). These workers have examined growth patterns from a large number of different organisms and found that they normally were different enough to serve as a fingerprint identification. The bacteria were grown in brain-heart infusion media and the thermograms were recorded with a multichannel batch calorimeter. Similar results have also been reported by Staples et al. (26). Johansson et al. (27) have developed a computer based identification technique for thermograms of this kind.

Ljungholm et al (28) have recently concluded a microcalorimetric study on mycoplasmas. Also for these microorganisms growth patterns showed characteristic differences which may be used for identification purposes.

It seems likely that the growth pattern identification method can be developed to a powerful technique in clinical bacteriology. However, bacterial growth patterns are sensitive to variations in experimental conditions, not the least the oxygen potential (29), which thus has to be carefully controlled. It is believed that there is a great need for more thorough research concerning the biochemical and biological basis to the thermogram.

Effect of antibiotics. Bindford et al. (30) have described a flow calorimetric screening method for antibiotic sensitivity testing. A culture of growing bacteria was established the increase in heat effect (being related to bacterial growth) was observed. In subsequent experiments solutions of antibiotica were added to new samples of the bacterial suspension. The sensitivities of the culture to the antibiotic tested were recorded in terms of the continued increase in heat effect (indicating resistent organisms), or unchanging or decreasing heat effect (indicating that the bacteria were sensitive to the drug tested). Good agreement was obtained with the disk agar diffusion method, but the calorimetric method provided results 12-24 h before the conventional technique.

Mårdh et al. (31,32) have recently studied the kinetics of the actions of various antibiotics. A batch calorimetric method as well as a flow method was used. Substantial differences in the kinetics of action were found, e.g. doxycycline and tetracycline, which could not be demonstrated by conventional bacteriological techniques such as the determination of MIC (minimum inhibitory concentration). The results suggested that the techniques used can find a practical value in establishing optimum doses and dose intervals in antibiotic therapy.

The same type of experiments with antibiotics have also been conducted recently by Semenitz and Tiefenbrunner (33)(34).

Blood cells

During the last few years several groups have performed calorimetric investigations on blood cells. The object has usually been to investigate the potential use of calorimetric techniques as a diagnostic tool in clinical hematology. Bandmann et al. (35) have made measurements on whole blood from healthy subjects and made a comparison with results obtained on purified samples of erythrocytes, granulocytes, lymphocytes, thrombocytes and cellfree plasma. The sum of the values for the fractions was found to be in full agreement with the value found for whole blood.

Erythrocytes. Following the first study by Levin and Boyo (36) a substantial systematic work has now been made on the heat production in erythrocytes from healthy human subjects (37-43). In the recent studies by Monti and Wadsö (41,42) the effect of variation of several experimental parameters was investigated (cell preparation technique, suspension media, pH, temperature, glucose concentration, storage conditions and the calorimetric technique). Using the derived data for the variation in heat effect with experimental conditions actual experimental values could be corrected to a given standard state (pH 7.40, 37°C, column separation technique, storage time 1 h at 4°C etc.). Such standard values are more suitable for comparison purposes than the directly obtained experimental values, e.g. when comparing data for cells from patients with normal values.

In addition to studies on erythrocytes from normal subjects measurements have been performed for erythrocytes from patients with different kinds of anemia. Boyo and Ikomu-Kumm (43) found highly increased values in sickle cell erythrocytes. Monti and Wadsö (37) and Levin (38) have reported significantly increased levels in other anemias. Results of an investigation by Monti and Wadsö (44) on erythrocytes from patients with hyperthyriodism suggest that calorimetric methods may have a practical value for the control of medical treatment of this disease.

The pentose shunt is stimulated in erythrocytes by the presence of methylene blue (Mb) and it has therefore been thought that heat effect measurements on erythrocytes during Mb stimulation may in some cases increase the diagnostic value of a calorimetric measurement. Monti and Wadsö (45) have reported results from a methodological study where they observed an increase in the heat effect by a factor of 10 (phosphate buffer, Mb concentration = 10^{-4} M), cf. Levin (46) and Minakami and de Verdier (40). Monti (47) has recently shown that erythrocytes from patients with chronic uremia have significantly increased heat effect values both in the presence of Mb and under unstimulated conditions.

<u>Leucocytes.</u> Levin has reported results from flow calorimetric measurements on leucocyte mixtures (i.e. essentially granulocytes and lymphocytes) from normal persons and from patients (48,49,50). The measurements were found to be disturbed by the adhesiveness of the cells (the granulocytes) and Levin attempted to overcome this by introducing air bubbles in the liquid flow (49). Using a static calorimetric method Bandmann and Wadsö (51) have further investigated the problem with adhesiveness of granulocytes, cf. (35). Usually a very large heat effect was observed during the initial

phase of the calorimetric measurement, in particular if the cells were suspended in buffer. With plasma suspensions better (although not always satisfactory) steady states were observed.

Levin (49) has reported results from phagocytosis stimulation of leucocytes by latex particles and subsequently Levin and Thomasson (50) utilized a similar technique in an interesting clinical application on leucocytes from patients with systemic lupus erythematosus (SLE). In this case the stimulation was caused by homogenates from the patients own leucocytes. A good discrimination was obtained between samples from normal subjects and from SLE-patients.

In their work on whole blood and its main fractions Bandmann et al. (35) reported results from static calorimetric measurements on lymphocytes. Good steady state values were observed but the heat effect values were found to be very dependent on the cell concentration. Krakauer and Krakauer (52) have recently made a systematic study on the heat production by lymphocytes under normal conditions and during stimulation with antigen (DNP-BSA), mitogenic lectin (Con A) and in the presence of various inhibitors. In a more preliminary study Johns et al. have reported results from other immunological experiments with leucocytes (53).

Thrombocytes. Levin (48) has reported results from flow calorimetric measurements on thrombocytes and Bandmann et al. (35) obtained very similar results in measurements under static conditions. A more systematic study of heat production in thrombocytes under different experimental conditions has recently been finished (54). In the study by Bandmann et al. (35) it was noted that thrombocytes initially showed very large heat effects if the cells had been stored at low temperature for a short time (ca. $4^{o}C$ for 0.5 h). It was concluded that the effect was due to cell aggregation. Thrombocyte aggregation has been studied in some detail by Ross et al. (55) using a batch calorimetric technique. These workers noted very significant heat effects during (mild) agitation and dilution of the thrombocyte suspensions.

Other Investigations on Cellular Material

Boivinet et al.(56) have demonstrated the potential use of microcalorimetry for the determination of hormone activity using rat epidydimal fat tissue kept in a nutritional buffer in the calorimetric vessel. When a solution of insulin was added there was a substantial increase in the heat effect. From the quotient between the increased heat effect and the original heat effect from the fat tissue an activity value for the insulin preparation can be calculated. In a study with isolated pancreatic β-cells Gylfe and Hellman (57) measured the heat effect with and without the presence of glucose.

In a search for a viability test for transplant organs Engberg et al. (58) have reported results from an exploratory study on rat kidney biopsies. The results showed a parallel decrease in heat effect and oxygen consumption during storage.

CONCLUSION

The recent developments in microcalorimetry have led to a significant interest for the technique in the medical field. As yet calorimetry has not found any use as a routine analytical tool in clinical work but it has become established on the research level as an analytical tool and for thermodynamic studies. It is felt that with the use of calorimetry a range of new and medically interesting analyses can be developed, perhaps in particular on cellular material. It is clear that the methodological work has only begun. There is a need for extensive studies, on a very basic level, concerning the detailed interpretation of recorded heat effect values and thermograms.

REFERENCES

1. Spink, C. and Wadsö, I.: Calorimetry as an analytical tool in biochemistry and biology. In D. Glick (ed.), Methods in biochemical analysis, vol. 23, p. 1-159, Wiley-Science, New York (1976).
2. Wadsö, I.: Microcalorimeters. Quart. Rev. Biophysics, 3, 383-427 (1970).
3. Benzinger, T.H. and Kitzinger, C.: Microcalorimetry, new methods and objectives. In C.M. Herzfield (ed.) Temperature - its measurment and control in science and industry, Vol. 3. Part 3. Reinholds, New York (1963).
4. Brown, H.D., Evans, W.J. and Altschul, A.M.: Analysis by differential calorimetry of ATPase activity in potato apyrose and red blood cell ghosts. Life Sci. 3, 1487 (1964).
5. Monk, P. and Wadsö I.: Flow calorimetry as an analytical tool in biochemistry and related areas. Acta Chem. Scand. 23, 29-36 (1969).
6. Konicekova, J. and Wadsö, I.: Use of flow calorimetry for the determination of choline esterase activity and its inhibition by organophosphorous pesticides. Acta Chem. Scand. 25, 2360-2362 (1971).
7. Beezer, A.E. and Stubb, C.D.: Application of flow microcalorimetry to analytical problems. I. Determination of organophosphorous pesticides by inhibition of choline esterase. Talanta, 20, 27-31 (1973).
8. Beezer, A.E., Steenson, T.I. and Tyrell, H.J.V.: Application of flow-microcalorimetry to analytical problems. II. Urea-urease system. Talanta 21, 467-474 (1974).
9. Yourtee, D.M., Brown, H.D., Chattopadhyay, S.K., Phillips, D. and Evans, W.J.: Microcalorimetry applied to clinical enzyme measurements. Anal. Lett. 8, 41-49 (1975).
10. Rehah, N.N., Everse, J., Kaplan, N.O. and Berger, R.L.: Determination of the activity and concentration of immobilized and soluble enzymes by microcalorimetry. Anal. Biochem. 70, 381-386 (1976).
11. Goldberg, R.N., Prosen, E.J., Staples, B.R., Boyd, R.N. and Armstrong, G.T.: Heat measurements applied to biochemical analysis: Glucose in human serum. Anal. Biochem. 64, 68-73 (1975).

12. McGlothlin, C.D. and Jordan, J.: Enzymic enthalpimetry, a new approach to clinical analysis: glucose determination by hexokinase catalyzed phosphorylation. Anal. Chem. 47, 786-790 (1975).
13. McGlothlin, C.D. and Jordan, J.: Thermochemical determination of glucose in serum plasma and whole blood without prior deproteinization. Clin. Chem. 21, 741-745 (1975).
14. Johansson, Å., Lundberg, J., Mattiasson, B. and Mosbach, K.: The application of immobilized enzymes in flow microcalorimetry. Biochem. Biophys. Acta 304, 217-221 (1973).
15. Mosbach, K. and Danielsson, B.: An enzyme thermistor. Biochem. Biophys. Acta 364, 140-145 (1974).
16. Mosbach, K., Danielsson, B., Borgerud, A. and Scott, M.: Determination of heat changes in the proximity of immobilized enzymes with an enzyme thermistor and its use for the assay of metabolites. Biochem. Biophys. Acta 403, 256-265 (1975).
17. Mattiasson, B., Danielsson, B. and Mosbach, K.: Enzyme assay of cholesterol, glucose, lactose and uric acid in standard solutions as well as in biological samples. Anal. Letters, 9, 217-234 (1976).
18. Canning, L.M., Carr, P.W.: Rapid thermochemical analysis via immobilized enzyme reactors. Anal. Lett. 8, 359-367 (1975).
19. Schmidt, H.-L., Krisam, G. and Grenner, G.: Microcalorimetric methods for substrate determination in flow systems with immobilized enzymes. Biochim. Biophys. Acta 429, 283-290 (1976).
20. Bostic, W.D. and Carr, P.W.: A precise, continous recording clot timer based on a thermometric detection system. Amer. J. Clin. Pathol. 60, 330 (1973).
21. Watt, D., Berger, R.L., Green, D. and Marini, M.A.: Thermal titration: Application of calorimetry to the study of plasma coagulation. Clin. Chem. 20, 1013-1017 (1974).
22. Beezer, A.E., Bettelheim, K.A., Newell, R.D. and Stevens, J.: The diagnosis of bacteriuria by flow microcalorimetry. A preliminary investigation. Science Tools, 21, 13-16 (1974).
23. Kallings, L.O.: Report at Sec. Symp. on Microcalorimetry in Microbiology. Bedford College, London (1975).

24. Boling, E.A., Blanchard, G.C. and Russel, W.J.: Bacterial identification by microcalorimetry. Nature, 241, 472-473 (1973).

25. Russel, W.J., Farling, G.C., Blanchard, G.C. and Boling, E.A.: Interim review of microbial identification by microcalorimetry. In D. Schlessinger (ed.), Microbiology-1975, p. 22-31. American Society fo Microbiology, Washington, D.C. (1975).

26. Staples, B.R., Prosen, E.J. and Goldberg, R.N.: Fine structure in thermal growth patterns of bacteria by microcalorimetry. NBS report No 73-181, Washington, D.C., National Bureau of Standards (1973).

27. Johansson, Å., Nord, C-E. and Nordström, T.: Computer evaluation of bacterial growth patterns based on microcalorimetric data. Science Tools, 22, 19-21 (1975).

28. Ljungholm, K., Mårdh, P.-A., Renström, K. and Wadsö, I.: Microcalorimetric studies of mycoplasmas, acholeplasmas and ureaplasmas. J. Gen. Microbiol. In press (1976).

29. Monk, P. and Wadsö, I.: The use of microcalorimetry for bacterial classification. J. Applied Bact. 38, 71-74 (1975).

30. Binford, J.S., Binford, L.F. and Adler, P.: A semiautomated microcalorimetric method of antibiotic sensitivity tests. Am. J. Clin. Pathol. 59, 86-94 (1973).

31. Mårdh, P.-A., Ripa, T., Andersson, K.-E. and Wadsö, I.: Microcalorimetry as a tool for evaluation of antibacterial effects of doxycycline and tetracycline. Scand. J. Infect. Dis. In press (1976).

32. Mårdh, P.-A., Ripa, T., Andersson, K.-E. and Wadsö, I.: Kinetics of the actions of tetracyclines on E. coli as studied by microcalorimetry. Antimicrob. Ag. and Chemother. In press 10 (1976).

33. Semenitz, E. and Tiefenbrunner, T.: Mikrocalorimetrische Untersuchungen zur Bestimmung der Antibakteriellen Aktivität von Chemoterapeutika. Arzneimittelforschung. In press (1976).

34. Semenitz, E. and Tiefenbrunner, F.: Mikrokalorimetrische Untersuchungen zur Characterisierung des Wirkungstyps von Tetracyclinen. Med. Klinik. In press (1976).

35. Bandmann, U., Monti, M. and Wadsö, I.: Microcalorimetric measurements of heat production in whole blood and blood cells of normal persons. Scand. J. clin. Lab. Invest. 35, 121-127 (1975).

36. Levin, K. and Boyo, A.: Heat production from erythrocytes. Scand. J. clin. Lab. Invest. Suppl. 118, 55 (1971).

37. Monti, M. and Wadsö, I.: Microcalorimetric measurements of heat production in human erythrocytes. I. Normal subjects and anemic patients. Scand. J. clin. Lab. Invest. 32, 47-54 (1973).

38. Levin, K.: Determination of heat production from erythrocytes in normal man and in anemic patients with flow microcalorimetry. Scand. J. clin. Lab. Invest. 32, 55-65 (1973).

39. Levin, K., Fürst, P., Harris, R. and Hultman, E.: Heat production from human erythrocytes in relation to their metabolism of glucose and amino acids. Scand. J. clin. Lab. Invest. 34, 141-148 (1974).

40. Minakami, S. and de Verdier, C.-H.: Calorimetric study on human erythrocyte glycolysis. Heat production in various metabolic conditions. Eur. J. Biochem. 65, 451-460 (1976).

41. Monti, M. and Wadsö, I.: Microcalorimetric measurements of heat production in human erythrocytes. III. Influence of pH, temperature, glucose concentration, and storage conditions. Scand. J. clin. Lab. Invest. In press (1976).

42. Monti, M. and Wadsö, I.: Microcalorimetric measurements of heat production in human erythrocytes. IV. Comparison between different calorimetric techniques, suspension media and preparation methods. Scand. J. clin. Lab. Invest. In press (1976).

43. Boyo, A.E. and Ikomi-Kumm, J.A.: Increased metabolic heat production by erythrocytes in sickle-cell disease. Lancet 1, 1215 (1972).

44. Monti, M. and Wadsö, I.: Microcalorimetric measurements of heat production in human erythrocytes. II. Hyperthyroid patients before, during and after treatment. Acta Med. Scand. In press (1976).

45. Monti, M. and Wadsö, I.: Microcalorimetric measurements of heat production in human erythrocytes. Heat effect during methylene blue stimulation. Scand. J. clin. Lab. Invest. In press (1976).

46. Levin, K.: Microcalorimetric studies of human blood cells. Scand. J. clin. Lab. Invest. Suppl. 135, 32 (1973).

47. Monti, M.: Microcalorimetric measurements of heat production in erythrocytes of patients with chronic uremia. Scand. J. Haematol. Submitted (1976).

48. Levin, K.: Heat production by leucocytes and thrombocytes measured with a flow microcalorimeter in normal man and during thyroid dysfunction. Clin. chim. Acta 32, 87-94 (1971).

49. Levin, K.: A modified flow calorimeter adapted for the study of human leucocyte phagocytosis. Scand. J. clin. Lab. Invest. 32, 67-73 (1973).

50. Levin, K. and Thomasson, B.: Leucocyte stimulation by autohomogenate measured with flow microcalorimetry in Systemic Lupus Erythematosus. Acta Med. Scand. 195, 191-200 (1974).

51. Bandmann, U. and Wadsö, I.: Microcalorimetric measurements on human polymorphonuclear leucocytes. Scand. J. clin. Lab. Invest. Submitted (1976).

52. Krakauer, T. and Krakauer, H.: Antigenic stimulation of lymphocytes I. Calorimetric exploration of metabolic responses. Cell. Immunol. In press (1976).

53. Johns, P., Jasani, B. and Stanworth, D.R.: Microcalorimetry as a potential tool in the study of antibody antigen reaction systems incorporating a cellular element. J. Immun. Meth. 4, 83-116 (1974).

54. Monti, M. and Wadsö, I.: To be published.

55. Ross, P.D., Fletcher, A.P. and Jamieson, G.A.: Microcalorimetric study of isolated blood platelets in the presence of thrombin and other aggregating agents. Biochim. Biophys. Acta 313, 106-118 (1973).

56. Boivinet, P., Garrigues, J.C. and Grangretto, A.: Dosage microcalorimetrique de l'insuline. C.R. Soc. Biol. 162, 1770 (1968).

57. Gylfe, E. and Hellman, B.: The heat production of pancreatic β-cells stimulated by glucose. Acta Physiol. Scand. 48, 779-792 (1971).

58. Engberg, A., Frödin, L., Harvig, B. and Hellerström, C.: Potential use of microcalorimetry as a viability assay in renal preservation. Science Tools 22, 17-18 (1975).

5.2 Microcalorimetric Measurements of Heat Production in Human Erythrocytes with a Batch Calorimeter

M. Trumpa, B. Wendt

INTRODUCTION

The metabolism of human red blood cells (RBC) has been intensively studied by manometric (1) and enzymatic methods (1,2,3,4,6,7,8,9) and more recently by calorimetry as well (1,10,11,12,13,14,15). As the heat production by a single erythrocyte is very small ($Q = 8 \cdot 10^{-15}$ W) several milliliters of packed cells are necessary to obtain an appropriate signal. It has been observed that methylene blue (MB) stimulates metabolism and heat production up to six-fold. We want to report experiments on the stimulation of red blood cells from healthy donors and patients with various forms of anemia.

The main metabolic pathway for glucose in erythrocytes is glycolysis via the Embden-Meyerhof cycle. Under physiological conditions about 90 % of the total glucose is metabolized via glycolysis and about 10 % is converted via the pentose pathway into carbondioxid, which are also intermediate products of the glycolysis. The oxygen consumption of erythrocytes is very small, in the range of 0,3 mMol/l RBC (2,4). Most of the oxygen seems to be used for the oxidation of hemoglobin to methemoglobin (5,16,17).

In the early 1900's BARRON and HARROP discovered that an enormous increase in oxygen consumption can be caused by catalytic quantities of methylene blue (5). Methylene blue shifts the equilibrium between glycolysis and the pentose shunt in such a way that glucose is broken down completely by the pentose shunt (6).

METHODS AND MATERIAL

- All calorimetric experiments were performed with a microcalorimeter E. Calvet (SETARAM/LYON) with 4 vessels of 15 ml and a sensitivity of 61 μV/mW. The vessels were made of plexiglas to avoid possible iron poisoning from steel vessels. Usually, a 10 ml erythrocyte suspension, which had been passed through the SEC- Column (see below) were transferred to the vessels. The reaction temperature was 37 C.

- Manometric experiments were run in a Warburg Apparatus (BRAUN / MELSUNGEN) at the same temperature. The difference between respiration and fermentation could be determined by adding 0,2 ml of 20 % KOH to the manometric vessel.

- Red blood cells were prepared by filtration through a sulphoethylcellulose (SEC)* - Sephadex column. This filtration is based on ionic binding of white cells to SEC (18). The advantage of this method is that one does not lose the population of young erythrocytes, as normally occuring during washing procedures. Therefore, this population shows higher metabolic activity than old erythrocytes (13).

  * SEC (SERVA/HEIDELBERG)

- All calorimetric and manometric measurements were performed in buffer (pH 7,4, 37 C) containing 0,138 M NaCl, 0,005 M $MgCl_2$, 0,01 M KH_2PO_4 and 10 mM glucose. Methylene blue was used at a concentration of 0,1 mM.

RESULTS

a) Healthy donors

Figure 1 shows a typical thermogram of a red cell suspension incubated with methylene blue. The heat production of the red blood cells was calculated from the distance P_{tot} between baseline and heat flow line after two hours of incubation.

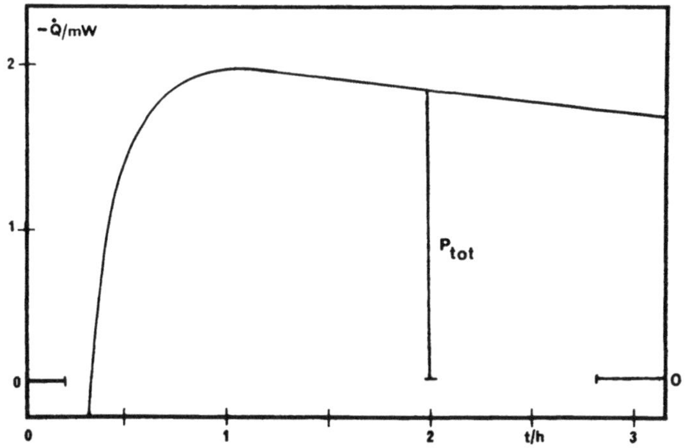

Figure 1: Thermogram of a red cell suspension with 0,1 mM MB in glucose-phosphate-buffer, 10 ml volume, number of erythrocytes: 3,7 · 10^9 /ml, temperature: 37 C, range 200 µV.

According to MONTI and WADSÖ (1975), the heat values of unstimulated erythrocytes were corrected to a standard pH of 7,40 with a correction factor of 1,3 % for a pH difference of 0,01 pH units, while the correction factor for stimulated cells was 0,4 % per 0,01 pH unit.

Table 1: Metabolic data from calorimetric measurements without methylene blue:

n = 6	$\bar{x}$ ± S.D.	Range	Dimension
Heat production	96 ± 18	69 to 129	mW/l RBC
Glucose consumption	1,8 ± 0,6	1,2 to 2,6	mM/l RBC·h
Lactate production	2,0 ± 0,3	1,7 to 2,5	mM/l RBC·h

Table 2: Metabolic data from calorimetric measurements with 0,1 mM methylene blue:

n = 33	$\bar{x}$ ± S.D.	Range	Dimension
Heat production	571 ± 64	367 to 720	mW/l RBC
Glucose consumption	2,9 ± 0,9	1,4 to 4,8	mM/l RBC·h
Lactate production	3,1 ± 0,5	2,1 to 3,8	mM/l RBC·h

MONTI and WADSÖ (14) reported a heat value of 102 ± 9 mW/l RBC for unstimulated red cells isolated with SEC and incubated in buffer and a corresponding glucose consumption of 1,8 ± 0,9 mM/l RBC·h.

Recent results of MINAKAMI and DE VERDIER (1) show that 50 % of the heat production of unstimulated erythrocytes stems from the formation of lactate with an enthalpy change of 71 kJ/Mol lactate.

In our experiments the uncorrected heat production was 73 mW/l RBC. If one interprets 50 % of this figure as due to lactate production, one gets a value of 66 kJ/Mol lactate, which is equal with the value of MINAKAMI et al. (1) within the limits of uncertainty.

Some authors (12,15) assumed that the pentose shunt also contributes a significant part to the heat production of erythrocytes. According to RAPOPORT et al. (7) and LIONETTI (3) who found that 10 % of the total glucose is metabolized via the pentose shunt (0,18 mM/l RBC·h in this investigation, see table 1) one gets an enthalpy changeof 728 kJ/Mol glucose consumed via the pentose shunt.

Combining these two figures one should be able to interpret the increased heat production of methylene blue stimulated erythrocytes, when glucose is totally metaboliced via the pentose shunt. One gets a theoretical value of 644 mW/ l RBC, which agrees approximately with the value actually found, 571 mW/l RBC. MINAKAMI et al. (1) report a value of 670 kJ/Mol O_2 when stimulating with methylene blue, while we find 787 kJ/Mol O_2. Thus, one may conclude that the pentose shunt is responsible for a significant part of the heat production in unstimulated cells.

b) Anemic donors

Several authors (11,12,15) observed that erythrocytes from anemic patients show an increased heat production, which could be due to increased glycolytic rates. In the present investigation the erythrocytes from one sideropenic anemia and six other anemias from patients with tumors in the bonemarrow were measured calorimetrically under stimulation with methylene blue. From figure 2 it is evident that the heat effects measured on samples from patients with tumors are scattered, but that they are higher on the average than those from the control group. This could be explained by higher metabolic rates via the pentose shunt.

If one assumes that the tumor anemias represent a homogenous group, their heat production is calculated to be 728 ± 67 mW/l RBC in contrast to 571 ± 64 mW/l RBC for the normal group. This difference is highly significant, while the heat production of the stimulated erythrocytes from the patient with sideropenic anemia seems to be normal.

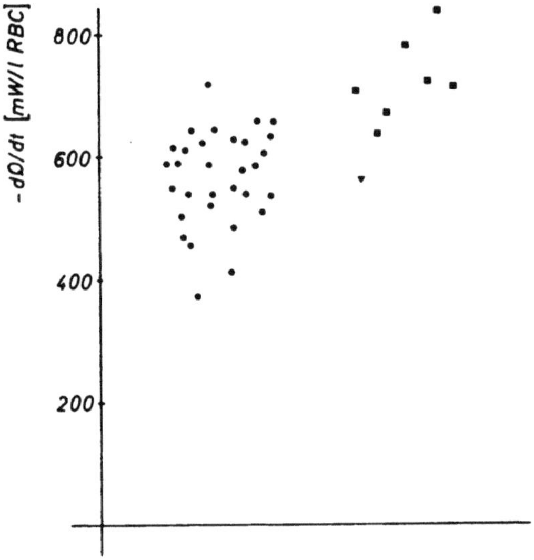

Figure 2: Heat production of stimulated red cells from normal and anemic patients: Normal red cells: • , sideropenic anemia: ▾ , tumor anemia: ■ , methylene blue concentration: 0,1 mM.

Finally it would be useful to investigate methylene blue stimulation in erythrocytes more thoroughly, to get more information about the activity of the pentose shunt under normal and pathological conditions.

Acknowledgement: The calorimeter was placed at our disposal by the Deutsche Forschungsgemeinschaft/Bonn-Bad Godesberg

REFERENCES

1) Minakami, S., de Verdier, C.H.: Calorimetric study on human erythrocyte glycolysis. Heat production in various metabolic conditions, Eur.J.Biochem. 65, 451-460 (1976)

2) Jacobasch, G., Minakami, S., Rapoport, S.M.: Glycolysis of the erythrocyte, in: Cellular and molecular biology of erythrocytes, Ed.: Yoshikawa, H., Rapoport, S.M., Urban und Schwarzenberg, München, Berlin, 55-92 (1974)

3) Lionetti, F.J.: Pentose phosphate pathway in human erythrocytes, in: Cellular and molecular biology of erythrocytes, Ed.: Yoshikawa, H., Rapoport, S.M., Urban und Schwarzenberg, München, Berlin, 143-166 (1974)

4) Huennekens, F.M., Liu, L., Myers, H.A.P., Gabrio, B.W.: Erythrocyte metabolism, III. Oxidation of glucose, J.Biol.Chem., 224-229 (1957)

5) Rapoport, S.M.: Medizinische Biochemie, VEB Verlag Volk und Gesundheit, Berlin (1975)

6) Rapoport, S.M.: Stoffwechselregulation der roten Blutzelle, Folia haematol. 83, 202-216 (1965)

7) Rapoport, S.M., Jacobasch, G.: Über die Regulation des Energiestoffwechsels des Erythrozyten. 1. Int.Symposium: "Stoffwechsel und Membranpermeabilität von Erythrozyten und Thrombozyten", Wien 17.-20. Juni 1968, Ed.: Deutsch, E., Gerlach, E., Moser, K., Thieme Verlag Stuttgart, 1-9 (1968)

8) Rapoport, S. M.: The regulation of glycolysis in mammalian erythrocytes in: Essays in biochemistry, Vol. 4, Ed.: Campbell, P. N., Greville, G. D., Academic Press, New York 69-103 (1968)

9) Rapoport, S. M.: Control mechanisms of red cell glycolysis, in: The human red cell in vitro, Ed.: Greenwalt, T. J., Jamieson, G. A., Grune and Stratton, New York, London, 153-178 (1973)

10) Bandmann, U., Monti, M., Wadsö, I.: Microcalorimetric measurements of heat production in whole blood and blood cells of normal persons, Scand. J. clin. Lab. Invest. <u>35</u>, 121-127 (1975)

11) Boyo, A. E., Ikomi-Kumm, J. A.: Metabolic activity of erythrocytes as determined by microcalorimetry, in: Protides of the biological fluids, Vol. 20, Ed.: Peeters, H., Pergamon Press Oxford, 559-562 (1974)

12) Monti, M., Wadsö, I.: Microcalorimetric measurements of heat production in human erythrocytes, I. Normal subjects and anemic patients, Scand. J. clin. Lab. Invest. <u>32</u>, 47-54 (1973)

13) Monti, M., Wadsö, I.: Microcalorimetric measurements of heat production in human erythrocytes. Comparison between different

calorimetric techniques, suspension media and preparation methods, Scand. J. clin. Lab. Invest., In press

14) Monti, M., Wadsö, I.: Report from a seminar in microcalorimetric investigations on human blood cells, unpublished (1975)

15) Levin, K.: Determination of heat production from erythrocytes in normal men and in anemic patients with flow-microcalorimetry, Scand. J. clin. Lab. Invest. 32, 55-65 (1973)

16) Kiese, M.: Die Reduktion des Hämiglobins, Biochem. Z. 316, 264-294 (1944)

17) Jaffé, E. R.: The formation and reduction of methemoglobin in human erythrocytes, in: Cellular and molecular biology of erythrocytes, Ed.: Yoshikawa, H., Rapoport S. M., Urban und Schwarzenberg, München, Berlin, 345-376 (1974)

18) Nakao, M., Nakayama, T., Kankura, T.: A new method for separation of human blood components, Nature New Biology 94, 246 (1973)

5.3 Characterization of the Mode of Action of Tetracyclines Using Microcalorimetry

E. Semenitz, F. Tiefenbrunner

The development of highly sensitive calorimeters made it possible to document microcalorimetric methods of investigation for the representation of heat production resulting through growth of microorganisms in liquid cultures (8, 14). In 1973 microcalorimetry was used for the representation of growth-inhibiting action of antibiotics on bacterial growth (1). In comparison to the above investigations, microcalorimetry was used for the characterization of the types of action of chemotherapeutics (17) and significant differences were found in the thermograms when ampicillin (a bactericidal antibiotic) respectively tetracycline (a bacteriostatic antibiotic) were added to growing bacteria in liquid culture media.

After a characteristic influence of ampicillin and tetracycline on the microcalorimetry curves of E. coli had been demonstrated (17), it was then investigated in the present study how the thermograms of a growing E. coli and Staph.aureus haem. culture are altered when different representatives of a single group of chemotherapeutics (for example, four tetracyclines: Tetracycline- HCl, Doxycycline, Minocycline and Demethyltetracycline) are added. In terms of antimicrobic activity these four tetracyclines are practically identical. However, Doxycycline and Minocycline still show a clearly visible growth-inhibition of different members of the species Staph.aureus haem. which are resistant to Tetracycline-HCl (2, 6, 7, 1o, 12, 13, 16, 18, 19). The molecular configurations, weights and formulae of these four tetracyclines are shown in Table 1.

TETRACYCLINE- HCl

$C_{22} H_{24} N_2 O_8 \cdot HCl$

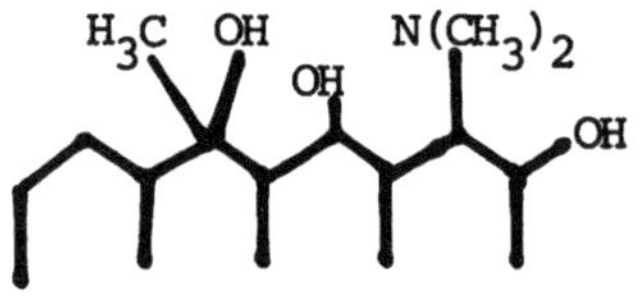

Molecular weight 48o

DEMETHYLCHLORTETRACYCLINE

$C_{21} H_{21} N_2 H_8 \cdot Cl$

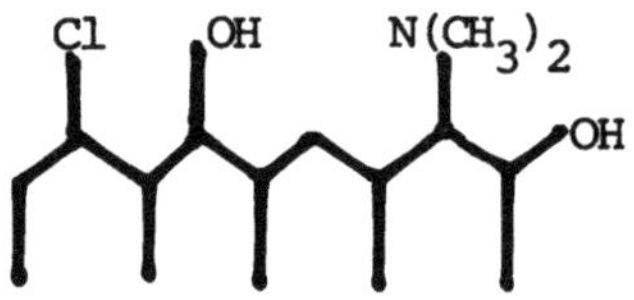

Molecular weight 491

DOXYCYCLIN

$C_{22} H_{24} N_2 O_8 \cdot H_2O$

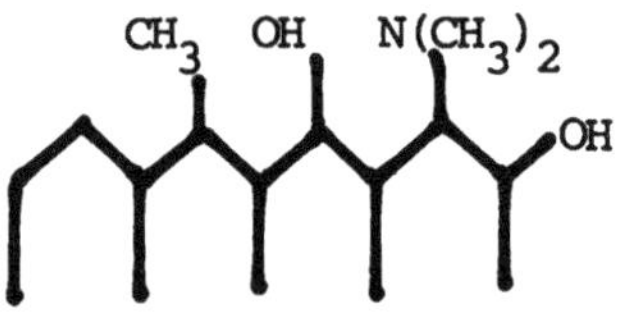

Molecular weight 462

MINOCYCLIN

$C_{23} H_{27} N_3 O_7 \cdot HCl$

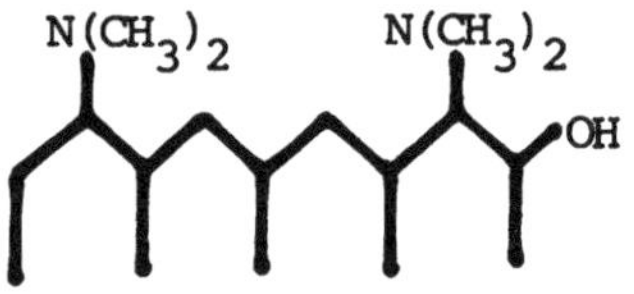

Molecular weight 493

Notably, these four tetracyclines can be differentiated according to their hydrophilic and lipophilic properties.

APPERATUS AND MODE OF TESTING

These investigations were carried out with a flow-microcalorimeter (LKB, Bromma, Type 21o7). A particular feature of this microcalorimeter is a 1 ml steel ampule measuring cell. The minimal detector sensitivity was o,o5 µV / µW; the minimal measurable continuous heat effect was about 1 µW. From a culture flask (5oo ml) situated in a water- thermostat (35°C) next to the microcalorimeter 2o ml/ h of growing culture were continuously stirred and pumped into the measuring cell of the calorimeter. For the experiments with E. coli as test organism Purple-Broth was used having 1% dextrose and for those with Staph. aureus haem. Columbia-Broth was used as culture medium. The test bacteria suspended in a physiol. NaCl-solution were added to the liquid culture media in such an amount that concentrations of $1o^6$ cells / ml of media were obtained at the beginning of the experiments.

TEST ORGANISMS

a) E. coli with haemolysis, H 3579, isolated from the urine of a patient suffering from cystitis. The minimal inhibition concentration (hereafter given as MIC) as determined by the micro inhibition test was 1.5 µg of each of the four tetracyclines per ml of culture medium.

b) Staph. aureus haem., A 3579. The MIC, similar to that of E. coli, H 3579, was identical with all four tetracyclines and measured o,7 µg of substance per ml of culture medium.

c) Staph. aureus haem., A 2953. The MIC of Tetracycline-HCl was 25o µg of substance per ml of culture medium, that of Demethyltetracycline 7o µg and that of Doxycycline and Minocycline identical 15 µg . Both Staph. species were isolated directly from puss samples.

RESULTS

Figure 1) shows the typical thermogram(=Normthermogramm,first thermogram on top of the figure) of E. coli, H 3579, in Purple-

Fig.(1)

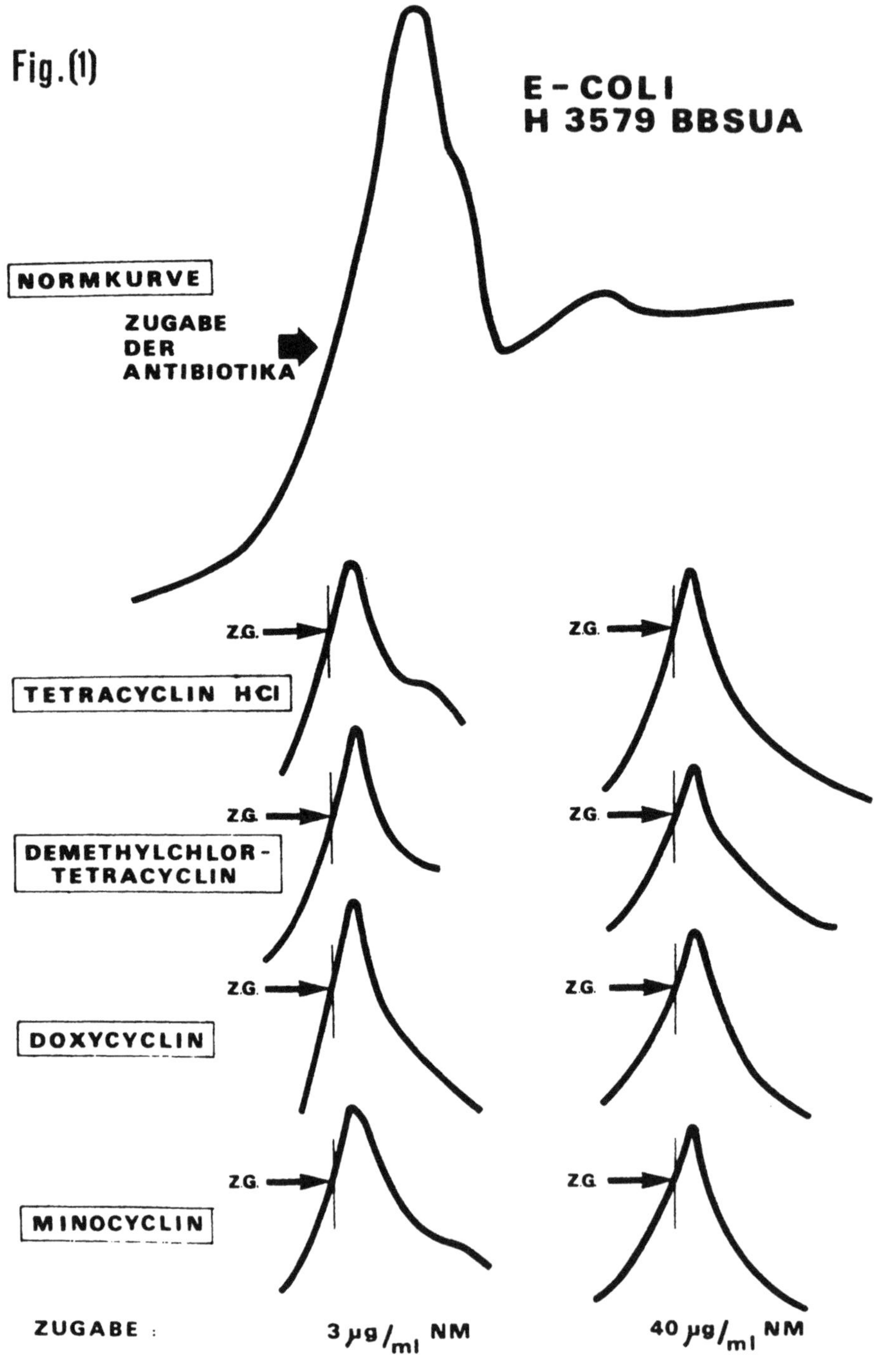
E-COLI
H 3579 BBSUA
NORMKURVE
ZUGABE
DER
ANTIBIOTIKA
Z.G.
TETRACYCLIN HCl
DEMETHYLCHLOR-
TETRACYCLIN
DOXYCYCLIN
MINOCYCLIN
ZUGABE :
3 µg/ml NM
40 µg/ml NM

Fig.(2)

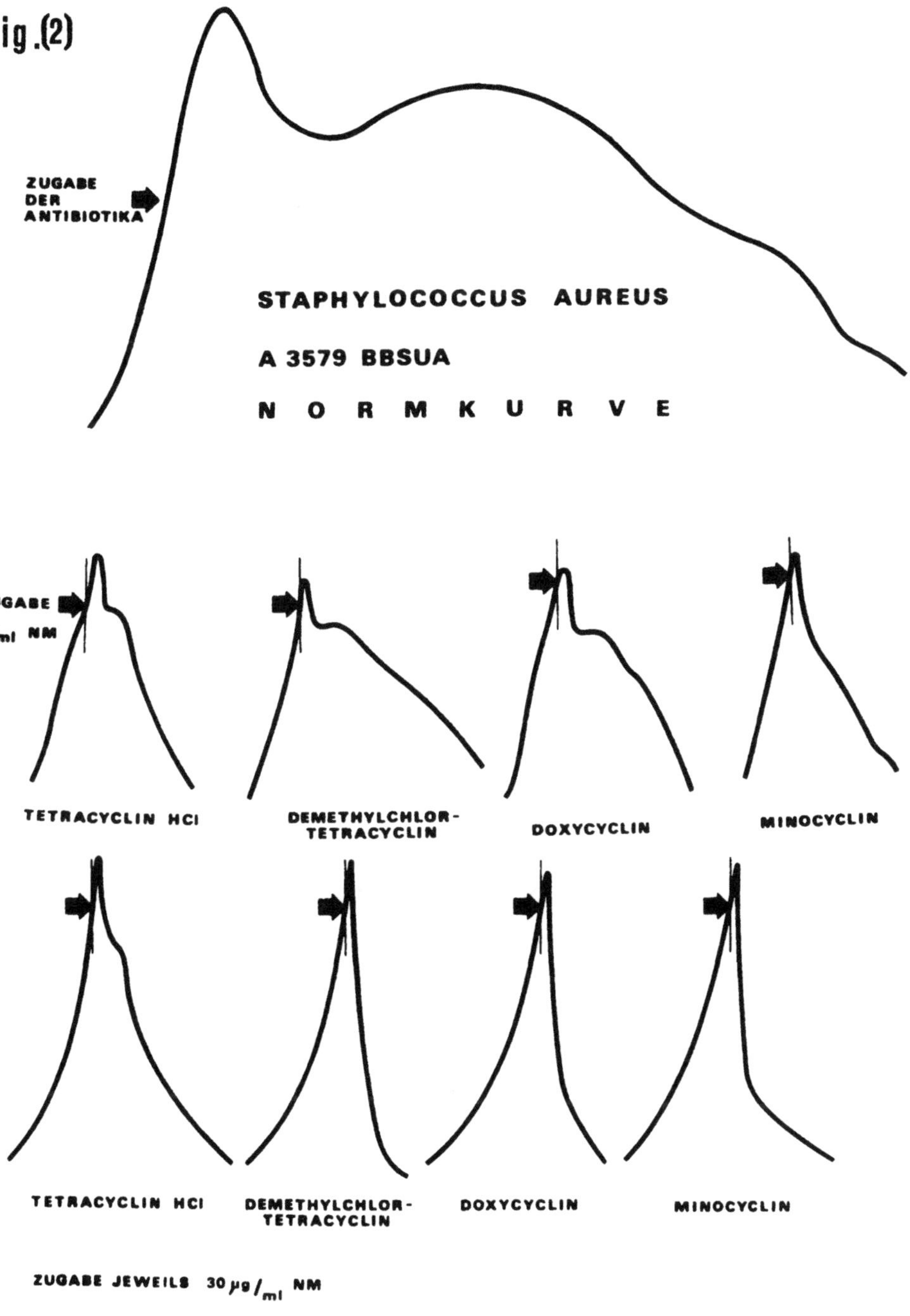
ZUGABE
DER
ANTIBIOTIKA
STAPHYLOCOCCUS AUREUS
A 3579 BBSUA
N O R M K U R V E
ZUGABE
3µg/ml NM
TETRACYCLIN HCl
DEMETHYLCHLOR-
TETRACYCLIN
DOXYCYCLIN
MINOCYCLIN
TETRACYCLIN HCl
DEMETHYLCHLOR-
TETRACYCLIN
DOXYCYCLIN
MINOCYCLIN
ZUGABE JEWEILS 30µg/ml NM

Fig.(3)

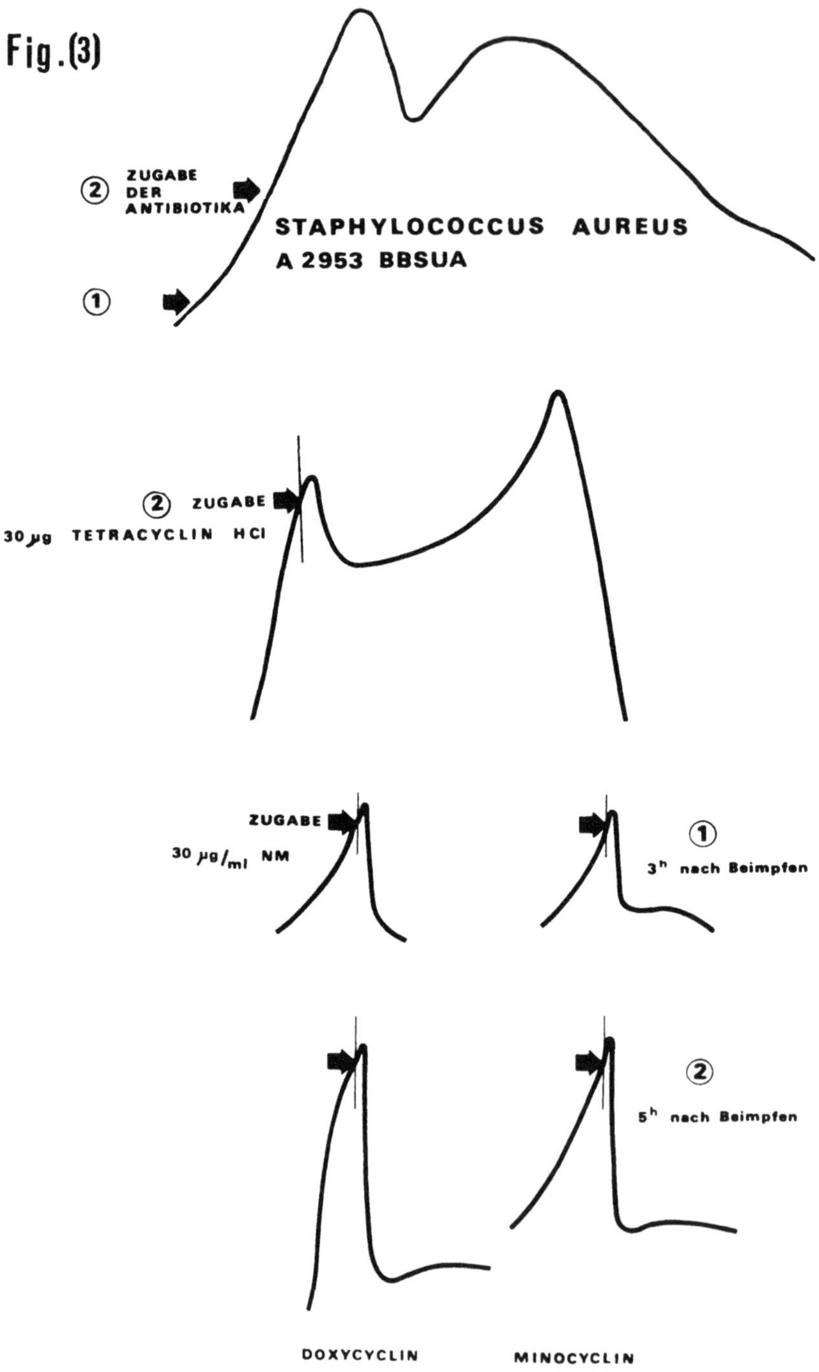
② ZUGABE DER ANTIBIOTIKA
STAPHYLOCOCCUS AUREUS
A 2953 BBSUA
①
② ZUGABE
30 µg TETRACYCLIN HCl
ZUGABE
30 µg/ml NM
①
3h nach Beimpfen
②
5h nach Beimpfen
DOXYCYCLIN
MINOCYCLIN

Broth. An arrow on the typical thermogram indicates the point at which in successive experiments the tetracyclines were added to the growing culture. This point was approximately 1.5 hours after the beginning of a measurable heat production, situated on the steeply rising section of the thermogram.

All four tetracyclines using a concentration of 3 µg and 4o µg of substance per ml of culture medium influenced the thermogram of E. coli identically (the MIC in the micro inhibition test was 1,5 µg /ml of culture medium). Figure 2 (top) shows the typical curve of Staph. aureus haem.,A3579, (the MIC was o,7 µg/ml). The addition of 3µg/ml of each tetracycline caused an immediate inhibition of heat production and the curves were identical. This was even more apparent adding 3o µg/ml .

Figure 3 shows the heat-time-curves of Staph. aureus haem., A 2953, i.e. of that Staph. aureus haem. which indicated different minimal inhibition concentrations. With the addition of 3o µg/ml of Tetracycline-HCl to the growing Staph. aureus, A 2953, culture there is a short stop in heat production followed by a second peak identical in altitude to that of the typical thermogram. The addition of 3o µg/ml of Doxycycline and Minocycline shows, however, practically the same inhibiting influence upon these growing cultures, as demonstrated in figure 2 for the more sensitive Staph. aureus A 3579.

DISCUSSION

It was determined that the addition of these four tetracyclines to the growing cultures of test bacteria a)(E. coli, H 3579) and b) (Staph. aureus haem.,A 2953), respectively those having uniform MIC`s of o,7 to 1,5 µg/ml , caused immediate inhibition of heat production . The minimal deviations in the curves as shown in figure 2) (using a 3 µg/ml concentr.)

were no longer detectable with the 3o µg/ml concentration. It is remarkable that the addition of only 3o µg/ml of Tetracycline-HCl to Staph. aureus haem.,A 2953, showing a MIC of 25o µg/ml in the micro inhibition test caused a clearly visible Stop in the heat production , even though followed by a further rise of the heat-time curve. Since the thermograms of both the resistante and the sensitive strain of Staph. aureus haem. were identical after addition of Doxycycline and Minocycline , there is no reason to assume an altered mechanism of action of these two antibiotics upon a tetracycline sensitive and a tetracycline resistant Staph. aureus haem. strain. It is generally believed that the tetracyclines inhibit protein synthesis in cell-free ribosome fractions. Their antibacterial activity is based on the concentration within the bacterial cell by an active uptake-mechanism (5, 15). This has been shown by E. coli as well as Staph. aureus species.

In kinetic studies with radioactive labelled tetracyclines it could be shown that E. coli assimilates tetracyclines in the range o,1 to 4oo µg without showing a saturation function. In the case of Staph. aureus using the same method a completely different uptake function, namely a two-phase function of the uptake system, was determined.

Using a fluorescence method an uptake for tetracyclines was found in Staph. aureus which corresponds within a concentration range of o,2 to 2oo µg to the mechanisms of saturation kinetics (3, 4, 5). These uptake studies with fluorescence methods are important for this particular investigation because it has been shown that a dependence of characteristic temperature ranges for the uptake of tetracyclines is present, corresponding to a specific configuration of fatty acids in the cell membrane. Differences in the effectiveness of lipophilic and hydrophilic tetracyclines could therefore be dependent upon the rate of penetration of these substances through the layer of fatty acids in the cell membrane (4) .

In summary, it can be said that microcalorimetry can be utilized to demonstrate that Doxycycline and Minocycline have the same mode of action as Tetracycline-HCl. The greater inhibiting effect of these two tetracyclines upon tetracycline- resistant gram positive cocci is not based on an altered uptake mechanism but on the fact that both of these substances are able to penetrate the cell membrane due to their lipohilic properties at a point when Tetracycline-HCl uptake is already blocked.

REFERENCES

(1) Binford, J.E.,Binford, L.F., and Adler,P.: A semiautomated microcalorimetric method of antibiotic sensitivity tests. Am. J. Clin. Pathol. 59, 86 - 94 (1973)

(2) Dimmling, Th.: Bakteriologische Untersuchungen über Resorption und Exkretion sowie über das Wirkungsspektrum von Alpha-6-Desoxyoxytetracyclin (Doxycyclin). Med. Klin. 62, 1269 - 1276 (1967)

(3) Dockter,M.E,. and Magnuson, J.A.:Kinetics of Chlorotetracycline uptake in Staphylococcus aureus by a fluorescence technique. Bioch. and Biophys. Res. Comm. 54, 79o - 795 (1973)

(4) Dockter, M.E.,and Magnuson, J.A.: Characterization of the active transport of Chlorotetracycline in Staphylococcus aureus by a fluorescence technique. J. Supramol. Struct. 2, 32 - 44 (1974)

(5) Dockter, M.E., and Magnuson, J.A.: Membrane Phase Transitions and the Transport of Chlorotetracycline. Arch. Bioch. and Biophys. 168, 81 - 88 (1975)

(6) Driessen, J.H.: A Computer Study of Bacterial Resistance Patterns to Antibiotics. Chemotherapy 21, (Suppl) 36 - 46 (1975)

(7) Fedorko, J., Katz, S., and Allnich, H.: In vitro activity of Minocyclin, a new tetracycline. J. Am. Med.

Scien. 255, 252 - 258 (1968)

(8) Forrest, W.W.: Microcalorimetry.In:Methods in Microbiology, ed. by Norris,J.R.,and Ribbons,D.W.,Academic Press Vol. 6B, 285 - 316 (1972)

(9) Isenberg, H.D.: In vitro activity of Doxycycline against bacteria from clinical material. Appl. Microbiol. 15, 1o74 - 1o78 (1967)

(1o) Klastersky, J., and Daneau,D.: Bacteriological Evaluation of Minocycline. Chemotherapy 17, 51 - 58 (1972)

(11) Mårdh, P.A.,Ripa,T.,Andersson,K.E.,and Wadsö,I.: Kinetics of the Action of Tetracyclines on Escherichia coli as Studied by Microcalorimetry.(1976) in press.

(12) Monnier,J.,Bourse,R.,and Onfray,J.: Doxycycline. Activité bactériostatique in vitro et taux sériques chez l' homme. Antibiotica 4, 268 - 282 (1966)

(13) Müller,J.: Die antibiotische Wirksamkeit von Doxycyclin, Tetracyclin und Chloramphenicol im Agardiffusionstest. Med. Klinik 65, 1821 - 1825 (197o)

(14) Prat, H.:Recent Progress in Microcalorimetry.(Ed. Scinner, H.A.),Pergamon Press, London (1963)

(15) Sazykin,Y.O., Fomina, I.V., Belyavskaya, N.S., Gryaznova, R.M., Petyshenko, R.M.: Studies on some Mechanisms of Resistance to Tetracyclines and their Transport to Bacterial Cell. Progress in Chemotherapy, Proceedings of the 8th Int. Congress of Chemotherapy, Athens, 1, 113 - 116 (1973)

(16) Semenitz, E.: Doxycyclin, Bakteriologische Untersuchungen. Wien. klin. Wochenschrft. 81, 454 - 457 (1969)

(17) Semenitz, E., and Tiefenbrunner, F.: Mikrokalorimetrische Untersuchungen zur Bestimmung der antibakteriellen Aktivität von Chemotherapeutika. Arzneimittelforschung, in press.

(18) Williamson, G.M.: Laboratory Studies on Doxycyclin. In International Symposium : New Resources in Antibiotic Therapy, Buenos Aires, June(1967) (typescript, 21 p.)

(19) Winkler, E., Weih, H.: Doxycyclin, Ergebnisse von Basisuntersuchungen. Med. Klinik 62, 1257 - 1262 (1967)

5.4 Whole Body Calorimetry

E. Jéquier

DIRECT AND INDIRECT CALORIMETRY

There are two main methods for measuring energy expenditure in man : direct and indirect calorimetry.

Considering the "heat balance equation" below, it can be seen that both methods do not measure the same energy : indirect calorimetry is used to determine the free energy produced in the body ($\dot{M}$), while direct calorimetry measures the rate of heat loss by the body ($\dot{R} + \dot{C} + \dot{E}$) :

(1) $\dot{M} = (\dot{R} + \dot{C} + \dot{E}) + \dot{S} + \dot{W}$

where $\dot{M}$ is the free energy production
$\dot{R}$ is the radiant heat exchange
$\dot{C}$ is the convective heat transfer
$\dot{E}$ is the evaporative heat transfer
$\dot{S}$ is the storage of body heat
$\dot{W}$ is the work done against external forces

Thus direct calorimetry can be used to measure metabolic rate providing that two conditions are met :

- whən body thermal equilibrium is reached, i.e. when body core and peripheral temperatures are constant ($\dot{S} = 0$), and,
- when no external work is done ($\dot{W} = 0$) .

Under these conditions, the heat balance equation is as follows :

(2) $\dot{M} = \dot{R} + \dot{C} + \dot{E}$

and direct calorimetry becomes a precise method for measuring the energy expenditure of a subject.

The principles of direct and indirect calorimetry are described below.

Direct human calorimeter

The total energy output of a subject is measured by using a human calorimeter or whole body calorimeter. It consists of a sealed insulated chamber in which the subject is placed. The modern instruments are heat flow calorimeters in which the rate of heat loss by a subject is measured with a gradient layer.

A direct calorimeter of low thermal inertia and rapid response time, large enough to accommodate patients, was built at the Division of Clinical Physiology in Lausanne (8). The principle of direct calorimetry for measuring heat flow by a gradient layer has been described by Benzinger and Kitzinger (1).

The heat flow through a layer is given by Fourrier's law :

(3) $\dot{Q} = A \frac{\lambda}{D} (T_1 - T_2)$

where $\dot{Q}$ is the flow through the layer, A is the area of the layer, λ is the specific thermal conductivity of the layer, and T_1 and T_2 are the temperatures of inner and outer surfaces, respectively.

The calorimeter is a small chamber with a volume of 1.56 m^3 (Fig. 1). The total inner surface is completely covered with a gradient layer which consists of an expoxy resin 2.4 mm thick with a copper circuit printed on both sides.

HUMAN DIRECT CALORIMETER

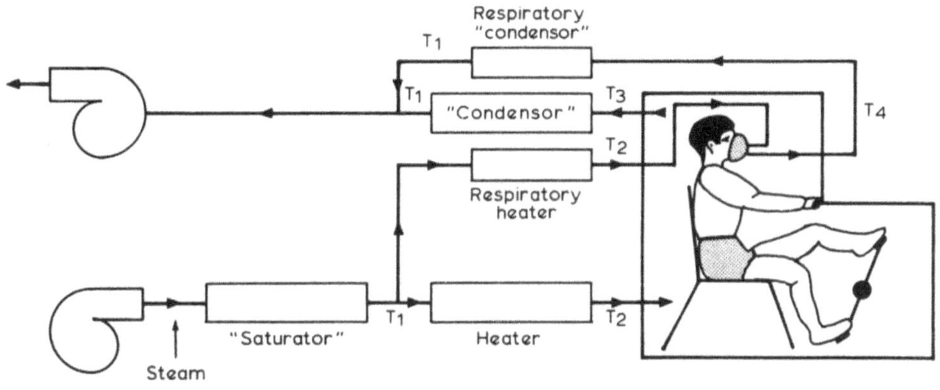

Fig. 1 Ventilatory circuit of the human direct calorimeter.

The electrical resistance of each circuit is dependent upon its temperature. The measurement of the heat flowing through the layer is performed by accurate measurements of the difference in temperature between the 2 surfaces of the layer. The copper circuits of the inner and outer surfaces of the gradient layer are connected in a Wheatstone bridge. The voltage measured is proportional to the sensible heat flowing through the layer ($\dot{Q}$). The measurement of detectable heat (radiative plus convective) using a gradient layer requires a constant temperature of the outer surface of the layer. The walls of the chamber are maintained at a constant temperature within 0.005°C of the set temperature by circulating water through channels in the walls of the chamber.

The evaporative heat loss of a subject is measured through a ventilatory circuit (Fig. 1). Outside air is forced through the ventilatory circuit at a flow rate of 1'150 l/min. The air is first cooled to a low temperature, T_1, and completely saturated with water. Before reaching the chamber, it is passed through a heater where it is warmed to the chamber temperature T_2, and relative humidity of the air decreases according to the new temperature T_2. The water vapor lost by the subject increases the water content of the air. Outgoing air from the chamber (at temperature T_3) passes through a heat exchanger (condenser) maintained at the low temperature, T_1. The water vapor lost by the subject condenses out and gives up its latent heat which is measured by a gradient layer in the condenser : this corresponds to the evaporative heat loss of the subject. Respiratory heat loss can be measured separately through a respiratory circuit.

Indirect calorimetry

Metabolic free energy production ($\dot{M}$) represents the rate of transformation of chemical energy into heat and mechanical work by aerobic and anaerobic processes within the organism. In steady state metabolic conditions, the total metabolic free energy production arises from oxidative metabolism, and the metabolic rate can be computed from the measurement of oxygen consumption ($\dot{V}_{O_2}$) using the caloric equivalent of a litre of oxygen, derived from the respiratory quotient.

In non-steady state conditions, at the onset of exercise, or after a change in work intensity, the rate of metabolic energy production ($\dot{M}$) exceeds that indicated by oxygen consumption. Part of the total energy production is due to aerobic processes which mainly result from the splitting of high-energy phosphate bonds of ATP and creatine phosphate. The amount of anaerobic energy production is usually evaluated by calculating the "oxygen deficit", which is obtained from the difference between the mean rate of oxygen consumption during the steady state, and the V_{O_2} actually measured during the "oxygen deficit" period. Although this method of evaluating the anaerobic energy production has been used by many investigators, it may yield partially erroneous results since the mechanical efficiency of the splitting of high energy phosphate bonds during anaerobic energy production may be different from the overall efficiency of aerobic metabolic processes.

Heat storage and changes in mean body temperature

The rate of heat storage ($\dot{S}$) is an important value for the study of thermoregulation in man. According to the heat balance equation :

(4) $\dot{S} = \dot{M} - (\dot{R} + \dot{C} + \dot{E} + \dot{W})$

$\dot{M}$ is obtained from oxygen consumption, $\dot{R} + \dot{C} + \dot{E}$ from direct calorimetry and $\dot{W}$ from measurement of the mean power of muscular exercise.

Integration of the rate of heat storage during a period of time $(t-t_o)$ gives the change in the body heat content (ΔS)

(5) $\Delta S = \int_{t_o}^{t} \dot{S}.dt$

The change in the body heat content is of great interest since recent works strongly suggest that the mammalian thermoregulatory system is mainly dependent upon information arising from many thermal receptors in the core and the periphery of the body. In other words, the main input to the thermoregulatory system is due to changes of the body heat content (ΔS).

From the measurement of ΔS we can calculate the change in mean body temperature ($\Delta \overline{T}_b$)

(6) $$\Delta \overline{T}_b = \frac{\Delta S}{m_b \, c_b}$$

where m_b = weight of the body (kg)
and c_b = specific heat of the body (3'475 $J.kg^{-1}.^{o}C^{-1}$)

Thermophysiologists usually calculate $\Delta \overline{T}_b$ using Burton's equation and weighting factors, as follows :

(7) $$\Delta \overline{T}_b = 0.9 \, \Delta T_{int} + 0.1 \, \Delta \overline{T}_{sk}$$

where T_{int} is the core temperature of the body, which is measured in the oesophagus or at the tympanic membrane, and $\overline{T}_{sk}$ is the mean skin temperature.

The weighting factors 0.9 and 0.1 are those proposed by Stolwijk and Hardy (9), but they cannot be considered as constant since they change with ambient conditions and during exercise. The calorimetric method for measuring ΔS and $\Delta \overline{T}_b$ can be used to assess the validity of the weighing factors used by thermophysiologists (4,5).

Cutaneous thermal conductance and cutaneous blood flow

Cutaneous thermal conductance ($\dot{K}_{sk}$) can be calculated according to Burton (3) and Hardy and DuBois (5) as follows :

(8) $$\dot{K}_{sk} = (\dot{R} + \dot{C} + \dot{E}_{sk}) / (T_{int} - \overline{T}_{sk}) \, A_b$$

where $\dot{E}_{sk}$ is the rate of evaporative heat loss from the skin:

(9) $$\dot{E}_{sk} = \dot{E} - \dot{E}_{ex}$$

$\dot{E}_{ex}$ represents the rate of respiratory heat loss and is often calculated from the equation of Mitchell et al. (7).

(10) $$\dot{E}_{ex} = 0.0023 \, (44 - P_a) \, \dot{M}$$

where $\dot{M}$, (the energy production), is expressed in Watt and

P_a (the ambient water vapor pressure) in mm Hg. $\dot{E}_{ex}$ can also be measured by direct calorimetry using a face mask to separate the heat loss by respiration from the skin heat losses (Fig. 1). A_b is the body surface area.

Cutaneous thermal conductance depends on the thermal conductivity of the tissues and of the amount of heat exchange due to convective heat transfer by the blood. Knowing the thermal conductivity of the tissues, thermal conductance measurements can be used to calculate the skin blood flow. Changes in cutaneous thermal conductance are due to variations in skin blood flow. The cutaneous blood flow ($\dot{F}_{sk}$), which is effective for heat exchanges, can be estimated by the following equation :

$$(11) \quad \dot{F}_{sk} = 60 \times 1000 \; \frac{\dot{K}_{sk} - \dot{K}_{tis}}{c_{bl} \; \rho_{bl}}$$

where $\dot{F}_{sk}$ is the cutaneous blood flow in $l.min^{-1}.m^{-2}$

$\dot{K}_{tis}$ is the tissue conductance (which is estimated to be 5.28 $W.m^{-2}.\ ^{o}C^{-1}$)

c_{bl} is the specific heat of blood (3'978 $J.kg^{-1}.{}^{o}C^{-1}$)

ρ_{bl} is the specific mass of blood (1'095.5 $kg.\ m^{-3}$)

The numeric factors transform the results into physiological units.

MEASUREMENTS OF DIRECT AND INDIRECT CALORIMETRY DURING EXERCISE

Simultaneous measurements of direct + indirect calorimetry, and of thermometry (skin temperature, tympanic temperature), were performed in 11 healthy untrained male subjects, whose age was 25 ± 1 year and weight 67.5 ± 2.5 kg. All subjects performed muscular exercise at ambient temperatures of 30°C and a relative humidity of 30 %. Results on skin blood flow are also presented for exercise performed at 20 and 25°C ambient temperatures.

The tests consisted of 4 periods :

1) 30 min of rest inside the calorimeter
2) 50 min of exercise at 40 W
3) 50 min of exercise at 90 W
4) 30 min of rest (recovery period)

Metabolic rate and heat losses

The results are shown in Fig. 2. During the rest period, the rate of metabolic heat production $\dot{M}$ was slightly above the rate of total heat losses ($\dot{H} = \dot{R} + \dot{C} + \dot{E}$). This indicates that the heat storage ($\dot{S}$) was slightly positive at 30°C ambient temperature.

At the beginning of exercise, we observed a weak increase in dry heat losses ($\dot{R} + \dot{C}$) due to movement of the legs. Evaporative heat losses ($\dot{E}$) (which included both cutaneous and respiratory heat losses) presented an exponential rise which was triggered at the beginning of the 40 W exercise period. The time constant of the evaporative heat losses (= time necessary to obtain 63.2 % of the response to a step change in heat load) was calculated to be in the order of 13.9 ± 2.2 min.

Heat storage

Total heat losses during exercise are obtained by adding the rate of work (horizontally hatched area) to the heat losses measured by direct calorimetry ($\dot{H}$). The difference between the rate of energy produced ($\dot{M}$) and the rate of energy losses ($\dot{H} + \dot{W}$) represents the rate of heat storage ($\dot{S}$), which is indicated by the obliquely hatched area. It is interesting to

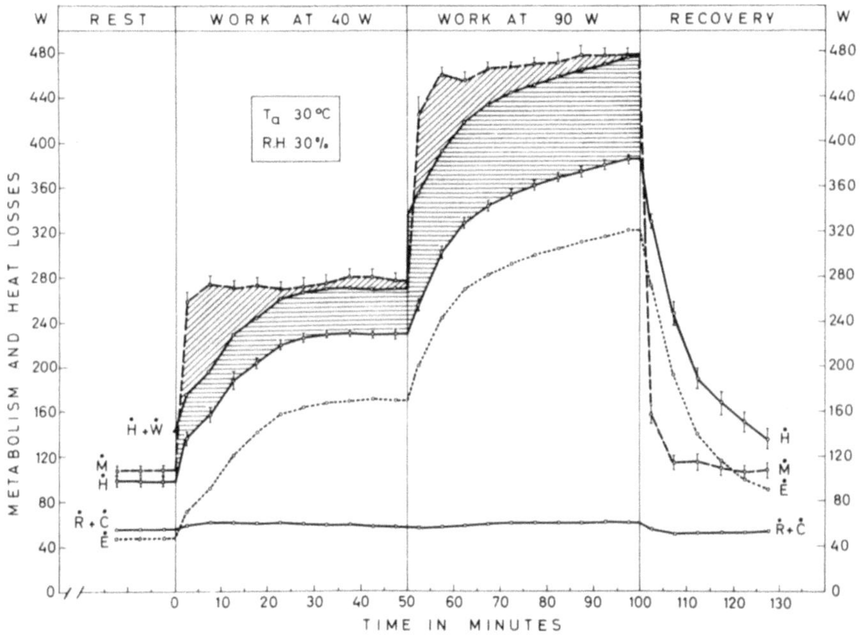

Fig. 2 Heat losses and metabolic rate during exercise at $30^{o}C$.

$\dot{E}$ = evaporative heat losses

$\dot{R}$ + $\dot{C}$ = radiative + convective heat losses

$\dot{M}$ = metabolic rate

Horizontally hatched area = work intensity

Obliquely hatched area = heat storage

(The anaerobic heat production is not shown on the figure).

(From : J. Appl. Physiol., 40, 384, 1976).

note that during exercise, the rate of heat storage tends towards zero indicating that a thermal equilibrium is reached, thanks to a precise regulation of evaporative heat losses.

At the beginning of the change in work intensity, a rapid additional increase in $\dot{E}$ occurs. Since the kinetics of the increase in $\dot{E}$ is slower than that of $\dot{M}$, a new heat storage ($\dot{S}$) is created. The time course of regulation of $\dot{S}$ is similar to that observed in the first exercise period (Fig. 2).

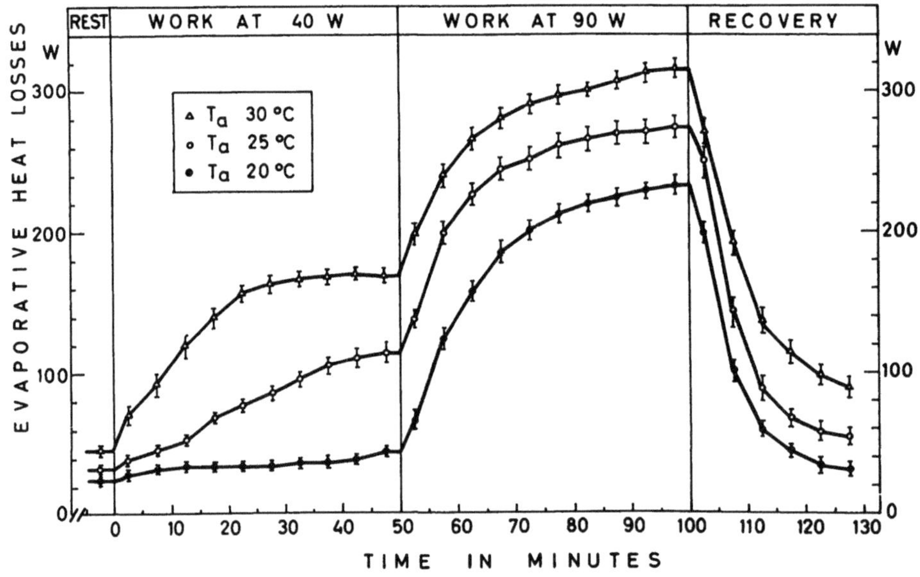

Fig. 3 Thermal conductance and cutaneous blood (effective for heat transfer) at rest and during exercise (40 and 90 W) at 20, 25 and 30°C.
(From : J. Appl. Physiol., 40, 384, 1976)

Cutaneous blood flow

Thermal cutaneous conductance ($\dot{K}_{sk}$) and cutaneous blood flow measured at 30°C increased significantly during the exercise periods. The increment in cutaneous blood flow is dependent upon the intensity of the exercise (Fig. 3). At ambient temperatures of 25 and 20°C, the increment in cutaneous blood flow was less marked than at 30°C.

THERMAL BODY INSULATION AND OBESITY

Thermal body insulation ($\dot{I}_b$) is the inverse of thermal conductance.

$$(12) \quad \dot{I}_b = \frac{1}{\dot{K}_{sk}} = \frac{(T_{int} - \overline{T}_{sk})\, A_b}{\dot{R} + \dot{C} + \dot{E}_{sk}}$$

Thermal body insulation depends on the composition and the thickness of subcutaneous tissue as well as on the vasomotor control of the cutaneous blood flow (6). The study of thermal body insulation in subjects of widely different body weight may give some insight into the mechanism of the apparent low energy expenditure observed in obese subjects exposed to cold or in anorexic patients exposed to a neutral ambient temperature.

Subjects : The subjects in this study were women with no evidence of endocrine diseases. They were assigned to 4 groups according to their weight and anamnesis.

Controls (n = 25) : women with body weight within ± 10 % of ideal weight.

Anorexic (n = 18) : women with body weight between -10 and -30 % of ideal weight and with a typical anamnesis of anorexia nervosa.

Trend to obesity (n = 29) : women moderately obese (weight + 10 to +35 % of ideal weight) and with an anamnesis of easy weight gain without obvious caloric excess.

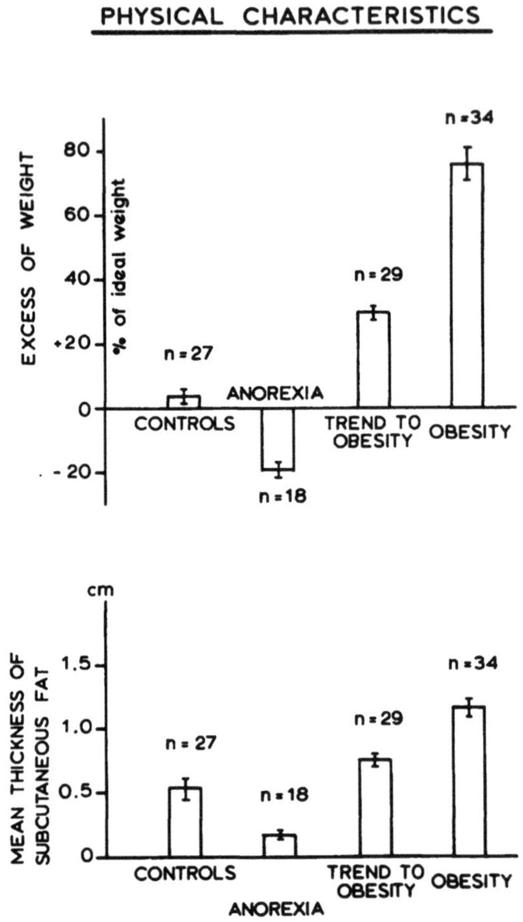

Fig. 4 Excess of weight (in percent of ideal weight) and mean thickness of subcutaneous fat in four groups of subjects.

<u>Obese</u> (n = 34) : women markedly obese whose body weight was in excess of +50 % (or more of ideal weight). The women were exposed to an ambient temperature of either 20°C or 28°C.

The mean weight and the mean thickness of subcutaneous fat of the 4 groups are shown in figure 4.

Heat losses

The figure 5 shows that controls had higher heat losses than the 3 other groups at 28 or 20°C.

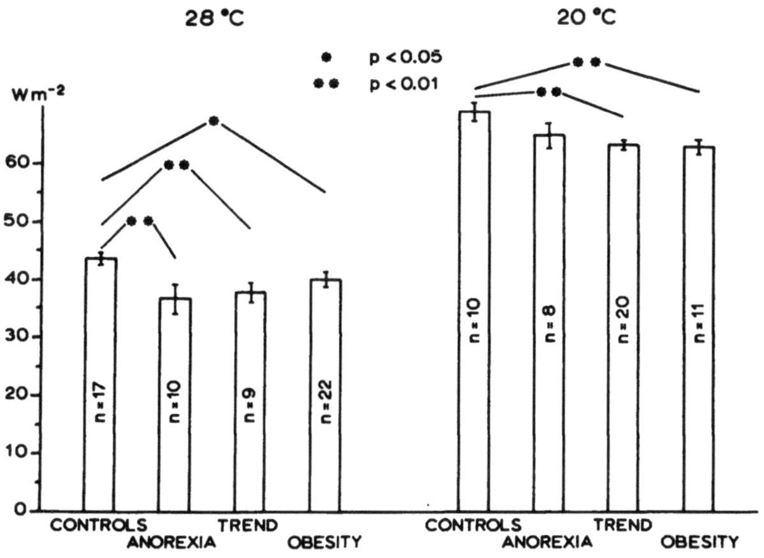

Fig. 5 Heat losses

Thermal body insulation

Thermal body insulation was obtained from calorimetry and thermometry as previously described. The insulation due to the subcutaneous fat was calculated from measurements of subcutaneous fat thickness and the coefficient of thermal conductivity of fat (0.21 W m^{-1} $^{o}C^{-1}$) as described by Jéquier et al. (6).

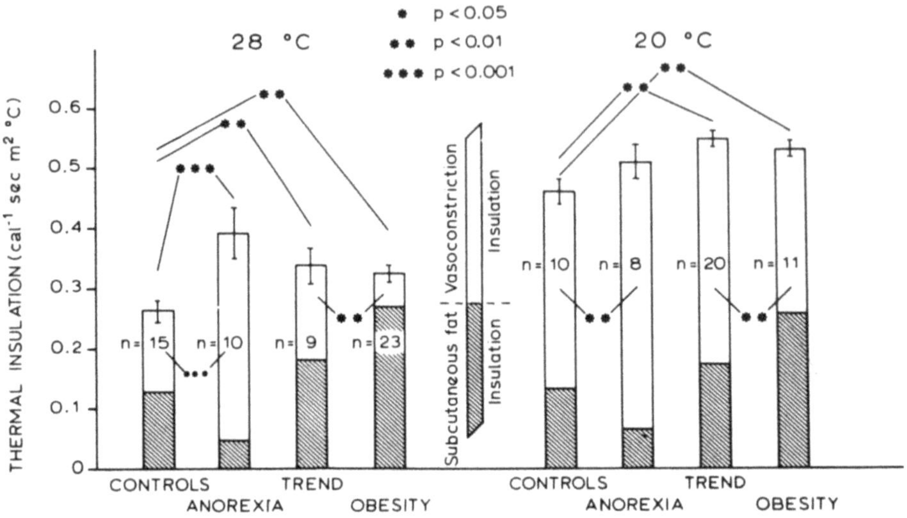

Fig. 6 Thermal insulation.

Figure 6 shows that the subcutaneous fat insulation depends upon the body weight of the subjects. The white area of the columns represents the insulation due to the decrease of blood flow by vasoconstriction in the muscular tissue under the subcutaneous fat. It is interesting to note that subjects with anorexia nervosa had the largest insulation at 28^{0}C; this was mainly due to decreased blood flow. By contrast, obese subjects had a large fat insulation and a small "regulatory" insulation. Both the trend to obesity group and the obese group showed a higher body insulation than controls.

During 1 hour of cold exposure at 20^{0}C a rise in body insulation occurred in the 4 groups of subjects. The increment of insulation was particularly marked in the trend to obesity and obese groups. This indicates that fat subjects have a better defense against heat losses due to cold exposure than lean subjects. It is of interest that subjects with a trend to obesity developed a similar body insulation to the frankly obese during cold exposure in spite of the marked difference in the degree of obesity between them. This is due to the fact that the large thermal insulation observed in subjects with a trend towards obesity results mainly from an increased vasoconstriction of subcutaneous vessels with a decreased transfer of heat by the blood flow.

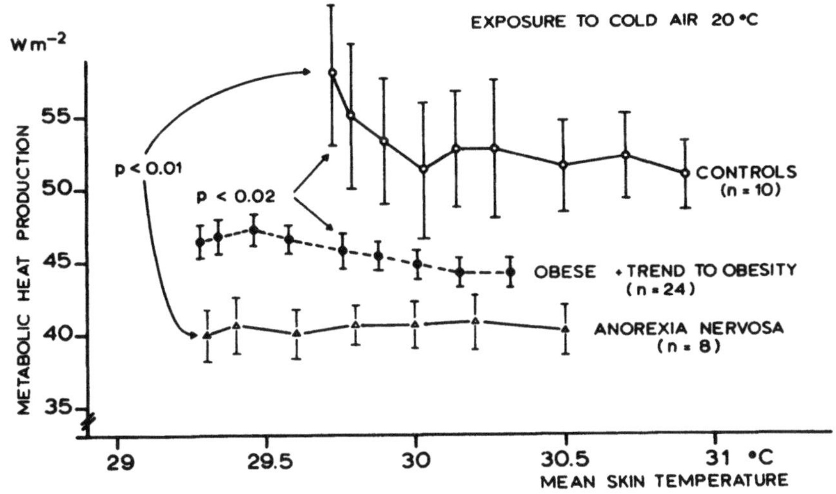

Fig. 7 Relationship between skin temperature and metabolic heat production.

Metabolic heat production

The physiological consequence of the large thermal insulation and of the decreased heat loss in both groups of obese and anorexic patients is a smaller metabolic response to cold. Since it is generally accepted that stimulation of thermogenesis due to cold exposure is related to the decrease in mean skin temperature ($\bar{T}_{sk}$) the metabolic heat production was plotted against $\bar{T}_{sk}$ (Fig. 7).

Figure 7 shows that the metabolic rate was lower in obese and anorexia nervosa subjects than in controls. In addition, control subjects showed a tendency to increase heat production (beginning of shivering) when mean skin temperature fell below 30°C. By contrast, no increase in metabolic rate was observed in either obese or anorexia nervosa patients. The skin temperature threshold at which metabolic rate was stimulated appeared to be lower in the latter than in controls. Moreover, obese, and to a lesser extend anorexia nervosa subjects, lose less heat per square meter of body surface area than controls. This economy of energy is probably related to the lack of stimulation of thermogenesis observed in these subjects during cold exposure.

THERMIC EFFECT OF NUTRIENTS

Another field of interest of whole body calorimetry is the study of the thermic effect of nutrients. The thermic effect consists of a rise in metabolic heat production following the ingestion of a meal.

The thermic effect of a glucose load was compared in two groups of subjects, obese and controls.

After a 50 g glucose oral load, the metabolic rate increased by 13.0 ± 1.5 KJ/m^2 in controls, and by 5.2 ± 1.3 KJ/m^2 in the obese (measurements performed during 150 min after the glucose intake.)

The changes in heat storage ($\dot{S}$) were measured by simultaneous measurements of direct and indirect calorimetry (Fig. 8). In the fasting state, controls presented a negative heat storage

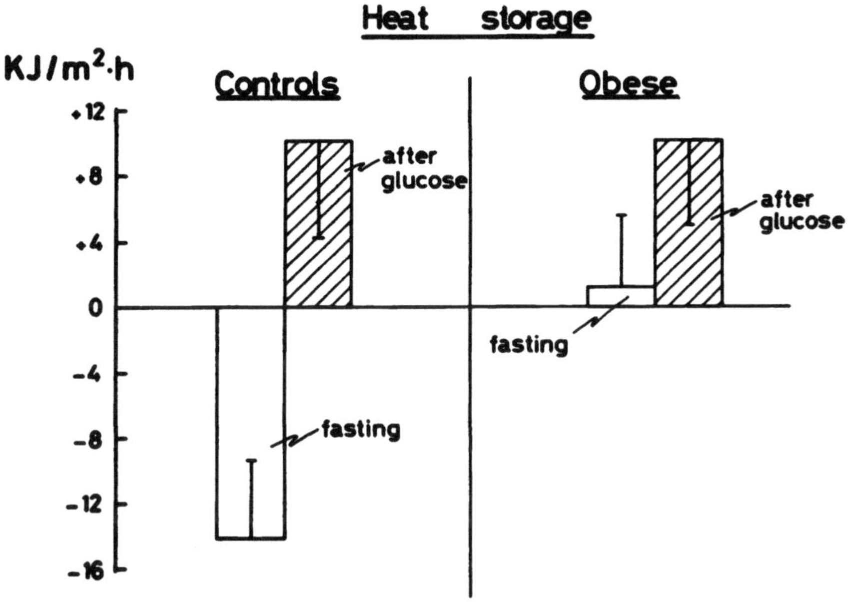

Fig. 8 Heat storage in the fasting state and after glucose ingestion.

while the obese were in thermal equilibrium ($\dot{S} \approx o$). After the 50 g glucose load, $\dot{S}$ changed markedly in controls (from -14 to +10 KJ/m^2.h), while $\dot{S}$ changed much less in the obese subjects (from +1 to +10 KJ/m^2.h).

The large thermic effect of glucose observed in control subjects has a thermoregulatory role since it changes the heat debt of the fasting state into a positive heat storage after ingestion of the nutrient. The fasting obese, being in thermal equilibrium, shows a smaller thermic effect after the glucose load. It remains to be determined whether the equilibrated thermal balance of the obese individual is involved in inhibiting the thermic effect found after glucose ingestion.

1. Benzinger, T.H. and Kitzinger, C. : Gradient layer calorimetry and human calorimetry. In : Temperature - Its Measurement and Control in Science and Industry. Edited by J.D. Hardy. New York : Reinhold, 1963, vol. 3, 87-109.

2. Bonjour, J.Ph., Welti, H.J. and Jéquier, E. : Etude calorimétrique des consignes thermorégulatrices au déclenchement de la sudation et au cours du cycle menstruel. J. Physiol., Paris, 72, 181-204 (1976).

3. Burton, A.C. : The application of the theory of heat flow to the study of energy metabolism. J. Nutr., 7, 497-533 (1934).

4. Gygax, P.H., Pittet, Ph. and Jéquier, E. : Dette de chaleur et isolation thermique chez la femme exposée au froid modéré : contribution de la calorimétrie directe à l'étude d'un modèle mathématique. J. Physiol., Paris, 72, 105-129 (1976).

5. Hardy, J.D. and DuBois, E.F. : Regulation of heat loss from the body. Proc. Natl. Acad. Sci., US 23, 624-631 (1937).

6. Jéquier, E., Gygax, P.H., Pittet, Ph. and Vannotti, A. : Increased thermal body insulation : relationship to the development of obesity. J. Appl. Physiol., 36, 674-678 (1974).

7. Mitchell, J.W., Nadel, E.R. and Stolwijk, J.A.J. : Respiratory weight losses during exercise. J. Appl. Physiol., 32, 474-476 (1972).

8. Spinnler, E., Jéquier, E., Favre, R., Dolivo, M. and Vannotti, A. : Human calorimeter with a new type of gradient layer. J. Appl. Physiol., 35, 158-165 (1973).

9. Stolwijk, J.A.J. and Hardy, J.D. : Partitional calorimetric studies of responses of man to thermal transients. J. Appl. Physiol., 21, 967-977 (1966).

6. Ecological Calorimetry

6.1 Thermodynamic Considerations of Invertebrate Anoxibiosis

E. Gnaiger

ECOLOGICAL AND BIOCHEMICAL BACKGROUND

In no case was it unequivocally demonstrated that any metazoan organism is capable of performing its whole life cycle under strictly anoxic conditions "(105)". There are, however, ecological indications of persistent, if not permanent, anoxic animal life, especially abundant in marine sediments containing hydrogen sulfide "(24,73,77,78,107,108,114,115)" and to a lesser extent in anoxic strata of limnetic muds "(3, 63, 84, 99)" and meromictic lakes "(4,90)".

Under these conditions glycogen is commonly the prime source of energy, whereas the metabolic integration of amino acids and lipids is still a matter of speculation and contradiction "(14,32,35,45,135)". Pathway and control of glucose degradation are comparable to the classical Embden-Meyerhof-Parnas route to the level of phosphoenol pyruvate (PEP) "(85,101)". A detailed literature deals with the aerobic-anoxic control of CO_2-fixation at the PEP-branchpoint and the formation of succinate via the reversed tricarboxylic-acid-cycle sequence (Fig.1). Most exentsively studied in parasitic helminths "(6,7,93,123)" and intertidal bivalves "(34,37,38,46,81, 135)" the same basic frame of the anoxic pathway leading to succinate was also demonstrated in fresh-water bivalves "(30)", gastropoda "(70,71,76,103,104)", polychaeta "(89,

121)", oligochaeta "(20,21,100,101)" and chelicerata "(23)". Additional studies indicate the existence of these metabolic pathways in other phyla "(32,39,41,50,67)", and support the view of its general importance in anoxibiotic invertebrates, and possibly in some vertebrates "(53,58,59,82)".

A principal criterion which must be fulfilled by any scheme of anoxic energy metabolism is the maintenance of redox balance, i.e. a constant NAD/NADH ratio. In lactic fermentation NADH, generated by the oxidation of glyceraldehyde-3-phosphate, is fed into the reductive lactate dehydrogenase step. As a peculiarity of molluscs, lactate dehydrogenase may be partially replaced by octopine dehydrogenase "(26,28,29,36, 87)". In animals with high anoxic capacity, however, the cytoplasmatic malate dehydrogenase takes over the function of lactate dehydrogenase and thereby plays an important role in the regulation of the PEP-branchpoint "(11,19,46,64,68, 69,112,122)" - quite analogous to the glycerol phosphate shuttle preventing aerobic formation of lactate (Pasteur effect).

Oxidation of an additional molecule of NADH proceeds concomitantly with the formation of succinate (Fig.1). As a mechanism sustaining a constant NAD/NADH ratio in Ascaris, Saz and his coworkers described a "malate dismutation system" giving rise to equal amounts of succinate and pyruvate by the combined avtivity of mitochondrial fumarate reductase (succinate dehydrogenase) and malic enzyme, respectively "(93,94,95,102)", but the NADH requirement along the metabolic path leading to other end products is not accounted for.

In molluscs the same dismutation system would be operative if succinate and alanine accumulate in equimolar amounts, alanine being formed from pyruvate by transamination reactions "(33,70,106)". Hochachka's group postulated that during anoxibiosis the tightly linked catabolism of carbohydrate and amino acids is responsible for the maintenance of redox equilibrium, whereby one glucose unit, two molecules of aspartate, and two molecules of α-ketoglutarate must be mobilized simultaneously, and end products must be formed in constant ratios "(45)". Experimental evidence, however, suggests anoxic formation of amino acids rather than amino acid consumption "(1,2,32,70,101,106,121,132,136)", an exception being Fasciola hepatica "(51,60)" a parasite exposed to a high-protein diet "(86)". This, together with the fact that anoxic end products do not accumulate at a fixed ratio (see below), supports the scheme given by de Zwaan and his colleges "(55,132,135,137)": By slow action in the forward direction the tricarboxylic acid cycle generates NADH which is consumed by the fumarate reductase step. Therefore anoxic carbohydrate and protein catabolism are not connected by an obligatory link, but, rather, are related to ecological and physiological parameters like quality of food, starvation, and time of anoxibiosis. This sets the stage for a numerical evaluation of anoxibiotic processes and comparison of different biochemical strategies of carbohydrate catabolism on quantitative grounds.

CHANGES OF FREE ENERGY ASSOCIATED WITH CATABOLISM OF GLUCOSE AND ANOXIC ATP GENERATION

The catalytic action of both pyruvate kinase in classical glycolysis and phosphoenolpyruvate carboxykinase "(110)" in the anoxic pathway, yields a net production of high-energy phosphate bonds by well known substrate level phosphorylation reactions. In the anoxic pathway, however, this is not the only ATP generating system. (The different nucleotide phosphates are treated as thermodynamically identical.) NADH-dependent phosphorylation by an anoxic electron transport chain with fumarate - analogous to oxygen - acting as terminal electron acceptor was demonstrated in parasitic hel-

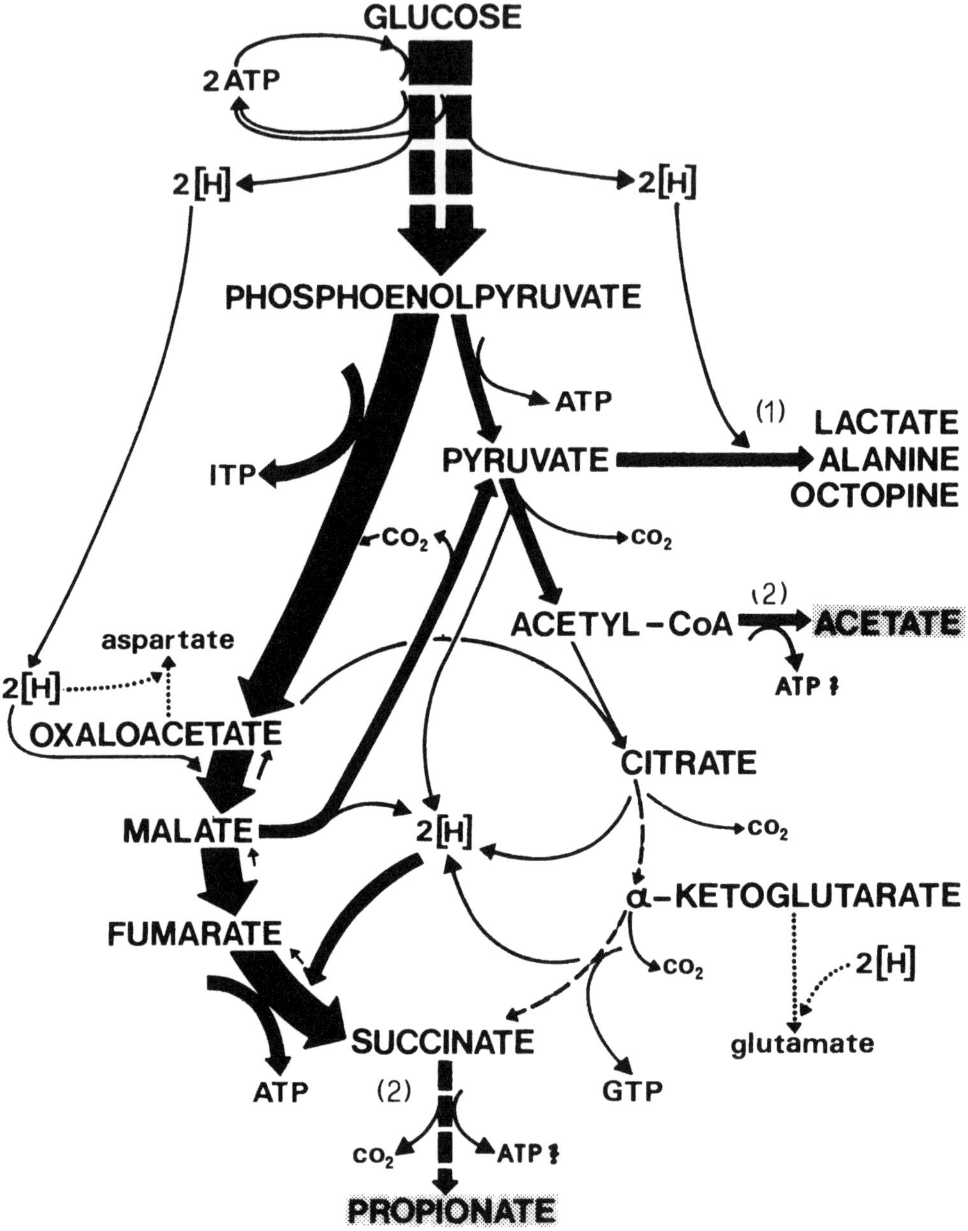

Fig.1. The basic frame of the anoxic pathways, and formation of primary (1) and secondary (2) end products in invertebrates.

minths "(15,16,17,18,57,98,102,124)" and seems to operate in other invertebrates "(40)" and in vertebrates as well "(43, 44,88,92,119)". As the reaction

$$(1) \quad \text{Fumarate}^{2-} + \text{NADH} + \text{H}^{+} = \text{Succinate}^{2-} + \text{NAD}^{+}$$

proceeds with $\Delta G^{o\prime}$ (change of free energy in standard state conditions at pH 7) of -67.7 kJ $mole^{-1}$ as calculated from standard tables "(66)", the free energy retained in ATP (30.5 kJ or 7.3 kcal $mole^{-1}$; "(62)") equals 45 %. The same thermodynamic efficiency is given for the aerobic electron transport chain (42 %;"(62)"). At "near physiological conditions"(see Tab.1) ΔG of reaction (1) amounts to -67.8 kJ $mole^{-1}$, and thermodynamic efficiencies are 70 % and 65 % for the anoxic and aerobic electron-transfer-coupled phosphorylations respectively.

Summarizing, half a glucose unit catabolized via the anoxic pathway (reaction 2) yields a net of 2 ATP. Redox balance is maintained if this reaction proceeds at a rate five times the rate of succinate formation via the tricarboxylic acid cycle (reaction 3): Two moles of phosphoenolpyruvate are converted into one mole of oxaloacetate and pyruvate respectively, giving rise to citrate by a condensation reaction (see Fig.1). In this way one mole of $succinate^{2-}$ (Suc) is derived from one mole of glucose (Glc), accompanied by three substrate level phosphorylation reactions.

$$(2) \quad 0.5\ \text{Glc} + \text{HCO}_3^- + \text{NADH} = \text{Suc} + 2\ \text{H}_2\text{O} + \text{NAD}^+$$

$$(3) \quad \text{Glc} + 4\ \text{H}_2\text{O} + 5\ \text{NAD}^+ = \text{Suc} + 2\ \text{HCO}_3^- + 5\ \text{NADH} + 9\ \text{H}^+$$

$$(4) \quad \text{Glc} + 0.86\ \text{HCO}_3^- = 1.71\ \text{Suc} + 1.71\ \text{H}_2\text{O} + 2.52\ \text{H}^+$$

A resultant reaction (4) proceeds with a net yield of 3.714 mole of ATP $mole^{-1}$ glucose. Compared to lactic fermentation the "succinic" fermentation produces 85 % more ATP per mole

<u>Tab.1.</u> **Changes of free energy and heat production in catabolism of glucose by anoxic pathways in invertebrates as compared with aerobic respiration.**

Metabolic Pathway	$-\Delta G^o$ (kJ)	$-\Delta G$ (kJ)	$-\Delta G$ (kcal)	ATP per Glc	Thermod. Efficiency	$-\Delta H$ (kJ)	$-\Delta H$ (kcal)	Caloric Efficiency
aerobic	2604	2903	693.9	36	58.6	2816	673	1.28
a) lactic	117	208	49.8	2	45.4	88	21	2.28
b) alaninic	140	209	50.0	2	45.2	599	143.2	0.33
c) succinic	171	273	65.3	3.7	64.2	255	60.9	1.46
d) **propionic**		(330)	(78.9)	3.7 - 5.4	53.2 - 77.7	181	43.6	3.00
e) acetic-succinic	204	288	68.9	3.3 - 4.0	54.1 - 65.6	240	57.3	1.67
f) **acetic-propionic**		(364)	(87.0)	3.3 - 5.3	43.3 - 69.3	183	43.8	2.89

Legend to Tab.1. Units are in kJ or kcal $mole^{-1}$ glucose; Temp.25°C; ΔG^0 values are for unit activity and CO_2 (aq); ΔG values are for 0.2 atm O_2, 0.05 atm CO_2 (aq), pH 7 and 0.01 M concentration of other reactants. Values of free energy are calculated from data according to Burton and Krebs "(12)" unless otherwise stated. ΔH values are calculated from heats of combustion "(5)" unless otherwise stated, and no corrections are made for heats of solution and dissociation of the acids. Thermodynamic efficiency is the percentage of ΔG retained in phosphorylation reactions (endergonic) per ΔG of the exergonic reaction. The caloric efficiency is the percentage of maximal ATP turnover in mole per kJ heat produced.

a) Glc = 2 $Lactate^-$ + 2 H^+

b) Glc + 2 $NH_4^+(aq)$ = 2 $Alanine^-$ + 2 H_2O + 4 H^+

Alanine production is assumed to proceed by NH_3 fixation onto pyruvate and therefore with maintenance of redox balance in the overall reaction. Transaminase reactions, however, proceeding without changes of free energy, may also play an important role. ΔH was calculated from heats of formation.

c) Glc + 0.857 HCO_3^- = 1.714 Suc^{2-} + 1.714 H_2O + 2.571 H^+

d) Glc = 1.714 Prp^- + 0.857 HCO_3^- + 2.571 H^+

Production of proprionate (Prp) from succinate (Suc) via the reversal of a reaction sequence known from catabolism of isoleucine and methionine was demonstrated in parasitic helminths "(96, 123)". Therefore the same considerations apply to "propionic" and "succinic" fermentation (see text), but an additional substrate level phosphorylation is probably driven by decarboxylation of succinate. As calculated from data of Wood "(120)", the biotin linked reaction

Methylalmonyl-CoA^- + H_2O = Propionyl-CoA + HCO_3^-

proceeds with ΔG = -33 kJ. As the methylmalonylmutase reaction and the transfer of CoA from propionate to succinate are readily reversible "(120)", the same value may be taken for approximating the overall decarboxylation of succinate to yield propionate.

e) Glc + 0.667 HCO_3^- =
= 0.667 Act^- + 1.333 Suc^{2-} + 1.333 H_2O + 2.667 H^+

f) Glc = 0.667 Act^- + 1.333 Prp^- + 0.667 HCO_3^- + 2.667 H^+

Formation of acetate delivers reducing equivalents for two fumarate reductase reactions. Therefore the NAD/NADH ratio is maintained, if acetate and succinate or propionate are produced in a ratio of 1 : 2. ATP production by acetate kinase is known from micro-organisms, but has not yet been proved in invertebrates.

of glucose, "propionic" fermentation probably 170 % more. In the latter case thermodynamic efficiency would approximate 80 %. Higher yields of ATP seem thermodynamically unfeasible. Increased efficiencies of phosphorylation in anoxic pathways are due not only to an increased thermodynamic efficiency, but also to larger increments of free energy in the respective reactions (Tab.1).

PRIMARY AND SECONDARY END PRODUCTS

The relative importance of the different anoxic pathways determines the overall efficiency of the anoxic energy metabolism. Due to short experimental acclimation periods (in the range of a few hours) the quantitative importance of lactate, alanine, and of succinate as end products of the anoxic energy metabolism has been largely overestimated. In these studies different sections were taken through a dynamic biochemical transitory process and described as different patterns of anoxic intermediary metabolism. This led to a confusing picture of various traits of metabolic strategies. Abundant evidence now exists that in invertebrates resistant to anoxia, lactate and alanine are initial end products only, accumulating in relatively small amounts during the first 12 to 24 hours of anoxibiosis "(10,27,55,113)". As time of anoxibiosis proceeds "propionic" and/or "acetic-propionic" fermentation become increasingly more important (Tab.2), "(30,56,135)", whereby in some cases the two volatile fatty acids are excreted after condensation to methylbutyrate "(21,97)".

Tab.2. Production of organic acids in µmol/g dry weight/h by *Tubifex* during successive periods of anoxic acclimation. The percentage of total organic acids is given in brackets. Calculated from data of Schöttler and Schroff "(101)".

hours	0 - 14	14 - 24	24 - 38	38 - 48
lactate	0.58 (4.2)	--	--	--
alanine	1.62 (11.8)	0.44 (4.8)	0.06 (0.6)	-0.06
succinate	4.33 (31.5)	3.33 (36.7)	2.48 (26.4)	1.20 (12.0)
acetate	1.64 (11.9)	1.90 (21.0)	2.36 (25.1)	2.30 (23.0)
propionate	5.57 (40.5)	3.40 (37.5)	4.50 (47.9)	6.50 (65.0)
glucosyl-equivalents [a)]	7.42	4.78	4.89	5.26
glycogen-consumption [b)]	6.29	5.59	4.77	3.52

a) The rate of consumption of glucosyl-equivalents is calculated from the rate of formation of end products under the assumption that glycogen is the sole source of energy and redox balance is maintained. The difference between this and the sum of organic acid production x 2^{-1} is proportional to the rate of the tricarboxylic acid cycle functioning in its forward direction.
b) Measured glycogen consumption in µmol glucosyl-equivalents/g dry weight/h.

The significance of lactate and alanine as primary end products of anoxibiosis is therefore seen in their influence on the activities of regulatory enzymes during the aerobic-anoxic transition of intermediary metabolism "(46,47,65,75,

74,116,130,131,132,133,134)". Accumulation of these primary end products, the formation of which is less efficient in generating high-energy phosphate bonds, initiates a new state of anoxi-metabolic equilibrium. Essentially, the observed trend can be interpreted as the general tendency to stabilize anoxic pathways with the highest yield of ATP (Fig.2).

APPLICATION OF DIRECT CALORIMETRIC METHODS

The confusing variety of species-specific patterns of organic acid production not only during anoxic acclimation, but in different tissues "(1,14,31,64)", developmental stages "(48, 91,113)", different sexes "(9)", and seasons "(121)" makes the biochemical assessment of the level of anoxic energy metabolism a tedious task. The variable ratio of accumulated and excreted end products "(22,72,101)" necessitates the consideration of both. Discontinuous carcass analysis is therefore unavoidable, whereby subtractions of estimated values measured on different individuals will introduce substantial errors due to considerable inhomogeneities, especially when large quantities of analytical substances (e.g.glycogen) are involved "(27,30,101,138,139)".

Heat production is an unspecific measure of the enthalpy changes accompanying all metabolic reactions of an organism. In long-term experiments continuous registration of heat production is possible, until a new steady state of anoxic energy metabolism - if ever - is reached. Turnover of ATP comprises the essential parameter of interest, since the scope of energy metabolism is the generation of high energy phosphate bonds as intermediate driving forces for all energy requiring biological processes. Provided that net synthesis is negligible during the experimental period "(49)", the caloric efficiency of ATP production may be calculated from the relative contribution of the different anoxic pathways

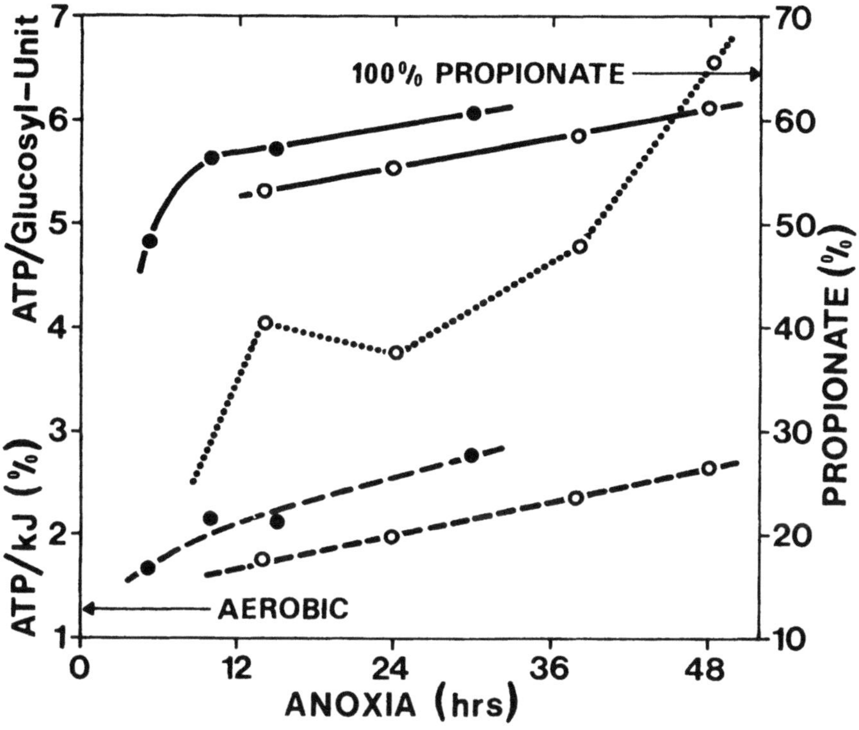

Fig.2. Anoxic acclimation of anoxibiotic invertebrates: Tubifex (open circles; calculated from data in Tab.2) and Anodonta (full circles; calculated from data of Gäde et al. "(30)". Solid line: Molar efficiency of ATP production in catabolism of glycogen. For every mole of glucose derived from glycogen one mole of ATP is saved, which is added to the maximal ATP production in the different anoxic pathways (Tab. 1). The arrow "100 % propionate" indicates the maximal molar efficiency of ATP production in pure "propionic" fermentation. Dashed line: Caloric efficiency. Values of Tab.1 are used in their proportional contribution to the anoxic metabolism without correcting for glycogen degradation. The calculated mean of total heat production is 85 μW/g dry weight of Tubifex and 97 μW/g dry weight of Anodonta. The arrow "aerobic" indicates the caloric efficiency level of aerobic metabolism. Dotted line: Percentage of propionate on total organic acid production.

to the overall metabolism (Tab.1). The biological interpretation of calorimetric data is difficult especially for the transitory period, but may be approximated by supplementary biochemical investigations (Fig.2). However, various side reactions "(25)", or energy production by "anoxic endogenous oxidation" "(42,54,125,126,127,128)" and net utilization of energy reserves "(109,117,118,129)" may obscure the calculated values.

Anoxic metabolism of diving vertebrates "(52,83)" and of whole benthic communities "(79,80)" was studied using direct calorimetric methods, but data of heat production of anoxibiotic invertebrates are not available so far. Inconsistencies of the presented calculations with future experimental investigations may reveal some gaps in our understanding of the biochemical mechanisms promoting anoxic animal life. Incorporating recent refinements of microcalorimetric techniques "(8,13,61,111)", the application of direct calorimetry to the study of ecological energetics will therefore contribute to better insight into the quantitative relationships of invertebrate anoxibiosis.

SUMMARY

New insight into the biochemical mechanism of invertebrate anoxibiosis made possible the calculation of the free-energy changes associated with the generation of high-energy bonds in nucleoside triphosphates (ATP, GTP, ITP) under anoxic conditions. The values obtained are compared with thermodynamic data of aerobic and fermentative energy production, and indicate a selection towards increased energetic efficiency of biochemical pathways leading to less toxic and readily excretable end products in anoxibiotic invertebrates. The thermodynamic model is mainly based upon a metabolic scheme elaborated on intertidal bivalves by de Zwaan et al. "(135)", benthic oligochaetes "(101)" and fresh-water bivalves "(30)". It may provide a general hypothesis for the energetic processes which operate in a variety of ecological and taxonomic groups of anoxibiotic animals.

This work was supported by the "Fonds zur Förderung der wissenschaftlichen Forschung in Österreich", project No.2919. I thank Prof.Dr.W.Wieser for discussions and reading the manuscript.

REFERENCES

1 Addink, A.D.F., Veenhof, P.R.: Regulation of mitochondrial matrix enzymes in Mytilus edulis L. Proc. 9th Europ.mar. biol. Symp., Aberdeen Univ.Press, 109-119 (1975).

2 Baginski, R.M., Pierce, S.K.: Anaerobiosis: a possible source of osmotic solute for high-salinity acclimation in marine molluscs. J.exp.Biol. 62, 589-598 (1975).

3 Beadle, L.C.: Adaptations of some aquatic animals to low oxygen levels and to anaerbic conditions. Symp.Soc.Exp. Biol. 15, 120-131 (1961).

4 Beadle, L.C.: Anaerobic life in a tropical crater lake. Nature 200, 1223-1224 (1963).

5 Blaxter, K.L.: The Energy Metabolism of Ruminants. Hutchinson Scientific and Technical, London. 3rd impr., 332 pp (1969).

6 Brand, Th.: Parasitenphysiologie. Gustav Fischer Verlag, Stuttgart, 353 pp (1972).

7 Brand, Th.: The biochemistry of helminths. Zeitschrift f. Parasitenkunde 45, 109-124 (1974).

8 Brown, H.D.: Biochemical Microcalorimetry. Academic Press, New York, 338 pp (1969).

9 Bruce, J.I., Ruff, M.D., Davidson, D.E., Crum, J.W.: Schistosoma mansoni and Schistosoma japonicum: Comparison of selected aspects of carbohydrate metabolism. Comp.Biochem.Physiol. 49B, 157-164 (1974).

10 Bryant, C., Morseth, D.J.: The metabolism of radioactive fumaric acid and some other substances by whole adult Echinococcus granulosus (Cestoda). Comp.Biochem.Physiol. 25, 541-546 (1968).

11 Bueding,E., Saz, H.J.: Pyruvate kinase and phosphoenolpyruvate carboxykinase activities of Ascaris muscle, Hymenolepis diminuta and Schistosoma mansoni. Comp.Biochem. Physiol. 24, 511-518 (1968).

12 Burton,K., Krebs, H.A.: The free-energy changes associated with the individual steps of the tricarboxylic acid cycle, glycolysis and alcoholic fermentation and with the hydrolysis of the pyrophosphate groups of adenosinetriphosphate. Biochem.J. 54, 94-107 (1953).

13 Calvet, E., Prat, H.: Recent progress in Microcalorimetry. Pergamon Press Oxford, 177 pp (1963)

14 Chaplin, A.E., Loxton, J.: Tissue differences in the response of the mussel Mytilus edulis to experimentally induced anaerobiosis. Biochem.Soc.Transact. 4, 437-441 (1976).

15 Cheah, K.S.: The oxidase systems of Moniezia expansa (Cestoda). Comp.Biochem.Physiol. 23, 277-302 (1967).

16 Cheah, K.S.: Properties of Ascaris muscle mitochondria. I. Cytochromes. BBA 387, 107-114 (1975).

17 Cheah, K.S., Bryant, C.: Studies on the electron transport system of Moniezia expansa (Cestoda). Comp.Biochem.Physiol. 19, 197-223 (1966).

18 Cheah, K.S., Chance, B.: The oxidase systems of Ascaris-muscle mitochondria. Biochem.biophys.Acta 223, 55-60 (1970).

19 Chen, C., Awapara, J.: Intracellular distribution of enzymes catalyzing succinate production from glucose in Rangia mantle. Comp.Biochem.Physiol. 30, 727-737 (1969).

20 Coles, G.C.: Modified carbohydrate metabolism in the tropical swamp worm Alma emini. Nature, Lond. 216, 685-686 (1967).

21 Coles, G.C.: Some biochemical adaptations of the swamp worm Alma emini to low oxygen levels in tropical swamps. Comp.Biochem.Physiol. 34, 481-489 (1970).

22 Dietz, T.H.: Body fluid composition and aereal oxygen consumption in the freshwater mussel, Ligumia subrostrata (Say): Effects of dehydration and anoxic stress. Biol.Bull. 147, 560-572 (1974).

23 Falkowski, P.G.: Facultative anaerobiosis in Limulus polyphemus: Phosphoenolpyruvate carboxykinase and heart activities. Comp.Biochem.Phsiol. 49B, 749-759 (1974).

24 Fenchel, T.M., Riedl, R.J.: The sulfide system: a new biotic community underneath the oxidiced layer of marine sand bottoms. Marine Biol. 7, 255-268 (1970).

25 Forrest, W.W., Walker, D.J., Hopgood, M.F.: Enthalpy changes associated with the lactic fermentation of glucose.J.Bacteriol. 82, 685-690 (1961).

26 Gäde, G.: Octopine dehydrogenase in the fresh water bivalve, Anodonta cygnea. Comp.Biochem.Physiol. 48B, 513-517 (1974).

27 Gäde, G.: Anaerobic metabolism of the common cockle, Cardium edule. 1.-The utilization of glycogen and accumulation of multiple end products. Arch.Internat.Physiol.Biochem. 83, 879-886 (1975).

28 Gäde, G.: Octopine dehydrogenase in the cockle Cardium edule. Biochem.Soc.Transact., 562nd meeting, Bangor 4, 433-436 (1976).

29 Gäde, G., Grieshaber, M.: Partial purification and properties of octopine dehydrogenase and the formation of octopine in Anodonta cygenea L. J.Comp.Physiol. 102, 149-158 (1975).

30 Gäde, G., Wilps, H., Kluytmans J.H.F.M., Zwaan, A.de: Glycogen degradation and end products of anaerobic metabolism in the fresh water bivalve Anodonta cygnea. J.Comp. Physiol. 104, 79-85 (1975).

31 Gäde, G., Zebe, E.: Über den Anaerobiosestoffwechsel von Molluskenmuskeln. J.Comp.Physiol. 85, 291-301 (1973).

32 Gilles, R.: Intermediary metabolism and energy production in some invertebrates. Archs.int.Physiol.Biochem. 78, 313-326 (1970).

33 Gilles, R.: Biochemical ecology of mollusca. Chemical Zoology Vol. 7, 467-499 (Florkin,M. and Scheer,B.T. eds.), Academic Press, New York (1972).

34 Goudsmit, E.M.: Carbohydrates and carbohydrate metabolism in mollusca. Chemical Zoology Vol. 7, 219-243 (Florkin, M. and Scheer,B.T. eds.), Academic Press, New York (1972).

35 Greichus, A. and Greichus, Y.A.: Lipid metabolism in the hog roundworm, Ascaris lumbricoides suum. Internat.J. Biochem. 6, 1-7 (1975).

36 Grieshaber, M., Gäde, G.: The biological role of octopine in the squid, Loligo vulgaris (Lamarck). J.Comp.Physiol. 108, 225-232 (1976).

37 Hammen, C.S.: Carbon dioxide fixation in marine invertebrates-V. Rate and pathway in the oyster. Comp.Biochem. Physiol. 17, 289-296 (1966).

38 Hammen, C.S.: Metabolism of the oyster, Crassostrea virginica. Am.Zool. 9, 309-318 (1969).

39 Hammen, C.S.: Lactate and succinate oxidoreductases in marine invertebrates. Mar.Biol. 4, 233-238 (1969).

40 Hammen, C.S.: Succinate and Lactate oxidoreductases of bivalve molluscs. Comp.Biochem.Physiol. 50B, 407-412 (1975).

41 Hammen, C.S., Osborne, P.J.: Carbon dioxide fixation in marine invertebrates: A survey of major phyla. Science 130, 1409-1411 (1959).

42 Harnisch, O.: Die Eigenart der oxybiotischen Erholung nach Anaerobiose bei Chironomidenlarven. Biol.Zentralbl. 69, 449-463 (1950).

43 Hoberman, J.D., Prosky, L.: The evidence of reduction of fumarate to succinate in perfused rat liver under conditions of reduced O_2 tension. Biochem.Biophys.Acta 148, 392-399 (1967).

44 Hoberman, H.D., Prosky, L., Arfin, H.W.: Reversal by anaerobiosis of the succinate dehydrogenase system in intact liver. Fedn.Proc. 24, 229 (1965).

45 Hochachka, P.W., Fields, J., Mustafa, T.: Animal life without oxygen: Basic biochemical mechanisms. Amer. Zool. 13, 543-555 (1973).

46 Hochachka, P.W., Mustafa, T.: Invertebrate facultative anaerobiosis. Science 178, 1056-1060 (1972).

47 Hochachka, P.W., Mustafa, T.: Enzymes in facultative anaerobiosis of molluscs-I. Malic enzyme of oyster adductor muscle. Comp.Biochem.Physiol. 45B, 625-637 (1973).

48 Holland, D.L., Tantanasiriwong, R., Hannant, P.J.: Biochemical composition and energy reserves in the larvae and adults of four British periwinkles Littorina littorea, L. saxatilis and L. neritoides. Marine Biol. 33, 235-239 (1975).

49 Hoogerheide, J.C.: Studies on the energy metabolism during anaerobic fermentation of glucose by baker's yeast. Radiation environmental Biophys. 11, 295-307 (1975)

50 Hulbert, W.C., Schneider, D.E., Moon, T.W.: Temperature and salinity adaptation in the purple shore crab Hemigrapsus nudus: An in vitro metabolic flux study with excised gills. Mar.Biol. 36, 223-231 (1976).

51 Isserhoff, H., Tunis, M., Read, C.P.: Changes in amino acids of bile in Fasciola hepatica infections. Comp. Biochem.Physiol. 41B, 157-163 (1972).

52 Jackson, D.C.: Metabolic depression and oxygen depletion in the diving turtle. J. Applied Physiol. 24, 503-509 (1968).

53 Johnston, I.A.: Anaerobic metabolism in the carp (Carassius carassius L.) Comp. Biochem.Physiol. 51B, 235-241 (1975).

54 Kerpel-Fronius, S., Zs.-Nagy, I.: Electron microscopic demonstration of energy production in molluscan neurons. Acta biol.Acad.Sci.Hung. 24, 83-90 (1973).

55 Kluytmans, J.H., Veenhof, P.R., Zwaan, A.de: Anaerobic production of volatile fatty acids in the sea mussel Mytilus edulis L. J.comp.Physiol. 104, 71-78 (1975).

56 Kluytmans, J.H., Zwaan, A.de: Production of volatile fatty acids in bivalved molluscs. Biochem.Soc.Transact. 4, 475-476 (1976).

57 Kmetec, E., Bueding, E.: Succinic and reduced diphosphopyridine nucleotide oxidase systems of Ascaris muscle. J.Biol.Chem. 236, 584-591 (1961).

58 Kornecki-Gerrety, E.H., Penney, D.G.: The effect of phosphoenolpyruvate, fructose-1,6-diphosphate, alanine and valine on pyruvate kinase from turtle liver and heart. Comp.Biochem.Physiol. 48B, 507-512 (1974).

59 Kornecki-Gerrity, E.H., Penney, D.G.: Kinetic properties of pyruvate kinases from turtle liver and heart. Comp. Biochem.Physiol. 49B, 15 (1974).

60 Kurelec, B.: Catabolic path of arginine and NAD regeneration in the parasite Fasciola hepatica. Comp.Biochem. Physiol. 51B, 151-156 (1975).

61 Lamprecht, I.: Biological calorimetry and DTA. Ecol. d'Ete de calorimetrie in Villeurbanne, Sept. 1973, 1-50 (Assoc Franc.de Calorimetrie et d'Anal.Thermique) P. Gravelle (ed.) (1973).

62 Lehninger, A.L.: Biochemistry. Worth Publ.,New York, 2nd ed., 1104 pp. (1975).

63 Lindeman, R.L : Experimental simulation of winter anaerobiosis in a senescent lake. Ecology 23, 1-13 (1942).

64 Livingstone, D.R.: Some kinetic and regulatory properties of the cytoplasmatic L-malate dehydrogenases from the posterior adductor muscle and mantle tissues of the common mussel Mytilus edulis. Biochem.Soc.Transact. 4, 447-451 (1976).

65 Livingstone, D.R., Bayne, B.L.: Pyruvate kinase from the mantle tissue of Mytilus edulis L. Comp.Biochem.Physiol. 48B, 481-497 (1974).

66 Long, C.: Biochemist's handbook. E. & F.N. Spon Ltd., London, 1192 pp. (1968).

67 Maguire, C., Boaden, P.J.S.: Energy and evolution in the thiobios: an extrapolation from the marine gastrotrich Thiodasys sterreri. Cahiers de Biol.Marine 16, 635-646 (1975).

68 McManus, D.P.: Tricarboxylic acid cycle enzymes in the plerocercoid of Ligula intestinalis (Cestoda: Pseudophyllidea). Zeitschr. Parasitenkunde 45, 319 (1975).

69 McManus, D.P., James, B.L.: Tricarboxylic acid cycle enzymes in the digestive gland of Littorina saxatilis rudis (Maton) and in the daughter sporocysts of Microphallus similis (Jäg.) (Digenea: Microphallidae). Comp.Biochem.Physiol. 50B, 491-495 (1975).

70 McManus, D.P., James, B.L.: Anaerobic glucose metabolism in the digestive gland of Littorina saxatilis rudis (Maton) and in the daughter sporocysts of Microphallus similis (Jäg.) (Digenea: Microphallidae). Comp.Biochem.Physiol. 51B, 293-297 (1975).

71 McManus, D.P., James, B.L.: Pyruvate kinases and carbon dioxide fixating enzymes in the digestive gland of Littorina saxatilis rudis (Maton) and in the daughter sporocysts of Microcephallus similis (Jäg.) (Digenea: Microphallidae). Comp.Biochem.Physiol. 51B, 299-306 (1975).

72 Mehlman, B., Brand, T.: Further studies on the anaerobic metabolism of some fresh water snails. Biol.Bull. 100, 199-205 (1951).

73 Moore, H.B.: The muds of the Clyde sea area. III. Chemical and physical conditions; rate and nature of sedimentation and fauna. J.mar.biol.Ass.U.K. 17, 325-358 (1931).

74 Mustafa, T., Hochachka, P.W.: Enzymes in facultative anaerobiosis of molluscs - II. Basic catalytic properties of phosphoenolpyruvate carboxykinase in oyster adductor muscle. Comp.Biochem.Physiol. 45B, 639-655 (1973).

75 Mustafa, T., Hochachka, P.W.: Enzymes in facultative anaerobiosis of molluscs - III. Phosphoenolpyruvate carboxykinase and its role in aerobic-anaerobic transition. Comp.Biochem.Physiol. 45B, 657-667 (1973).

76 Nopp, H.: Physiologische Aspekte des Trockenschlafes der Landschnecken. Sitzungsber.d.Österr.Akad.Wissensch., Mathem.-naturw.Kl.,Abt. 1, 182.Bd., 1-75 (1974).

77 Ott, J.A.: Determination of fauna boundaries of nematodes in an intertidal sand flat. Int.Rev.ges.Hydrobiol. 57, 645-663 (1972).

78 Ott, J., Schiemer, F.: Respiration and anaerobiosis of free living nematodes from marine and limnic sediments. Netherlands J. of Sea Research 7, 233-243 (1973).

79 Pamatmat, M.M.: In situ metabolism of benthic communities. Cahiers de Biol.Marine 16, 613-633 (1975).

80 Pamatmat, M.M., Bhagwat, A.M.: Anaerobic metabolism in lake Washington sediments. Limnol.Oceanogr. 18, 611-627 (1973).

81 Pandian, T.J.: Mechanisms of heterotrophy. Marine ecology, Vol.II, Part 1, 61-258 (O.Kinne (ed.), Wiley-Interscience, New York (1974).

82 Penney, D.G.: Effects of prolonged diving anoxia on the turtle Scripta elegans. Comp.Biochem.Physiol. 47A, 933-941 (1974).

83 Pickwell, G.V.: Energy metabolism in ducks during submergence asphyxia:assessment by a direct method. Comp. Biochem.Physiol. 27, 455-485 (1968).

84 Por, F.D., Masry, D.: Survival of a nematode and an oligochaete species in the anaerobic benthal of Lake Tiberias. Oikos 19, 388-391 (1968).

85 Prichard, R.K., Schofield, P.J.: The glycolytic pathway in adult liver fluke, Fasciola hepatica. Comp.Biochem.Physiol. 24, 697-710 (1968).

86 Prichard, R.K., Schofield, P.J.: The glyoxylate cycle, fructose-1,6-diphosphatase and glyconeogenesis in Fasciola hepatica. Comp.Biochem.Physiol. 29, 581-590 (1969).

87 Regnouf, F., Thoai van, N.: Octopine and lactate dehydrogenases in mollusc muscles. Comp.Biochem.Physiol. 32, 411-416 (1970).

88 Ringler, R.L., Singer, T.P.: Studies on the mitochondrial α-glycerophosphate dehydrogenase. J.Biol.Chem. 234, 8, 2211-2217 (1959).

89 Ruby, E.G., Fox, D.L.: Anaerobic respiration in the polychaete Euzonus (Thoracophelia) mucronata. Marine Biol. 35, 149-153 (1976).

90 Ruttner-Kolisko, A.: The vertical distribution of plankton rotifers in a small alpine lake with a sharp oxygen depletion (Lunzer Obersee). Verh.Internat.Verein.Limnol. 19, 1286-1294 (1975).

91 Salminen, K.: Succinate dehydrogenase and cytochrome oxidase in adult and plerocercoid Diphyllobothrium latum. Comp.Biochem.Physiol. 49B, 87 (1974).

92 Sanadi, D.R., Fluharty, A.L.: On the mechanism of oxidative phosphorylation. The energy-requiring reduction of NAD by succinate and the energy-yielding oxidation of NADH by fumarate. Biochemistry 2, 523-528 (1963).

93 Saz, H.J.: Facultative anaerobiosis in the invertebrates: pathways and controlsystems. Am.Zoologist 11, 125-135 (1971).

94 Saz, H.J., Hubbard, J.A.: The oxidative decarboxylation of malat by Ascaris lumbricoides. J.Biol.Chem. 225, 921-933 (1957).

95 Saz, H.J., Lescure, O.L.: The function of phosphoenolpyruvate carboxykinase and malic enzyme in the anaerobic formation of succinate by Ascaris lumbricoides. Comp.Biochem.Physiol. 30, 49-60 (1969).

96 Saz, H.J., Vidrine, A.: The mechanism of formation of succinate and propionate by Ascaris lumbricoides muscle. J.Biol.Chem. 234, 2001-2005 (1959).

97 Saz, H.J., Weil, A.J.: The mechanism of the formation of α-Methylbutyrate from carbohydrate by Ascaris lumbricoides muscle. J.Biol.Chem. 235, 914-919 (1960).

98 Scheibel, L.W., Saz, H.J., Bueding, E.: The anaerobic incorporation of ^{32}P into adenosine triphosphate by Hymenolepis diminuta. J.Biol.Chem. 243, 2229-2235 (1968).

99 Schiemer, F., Löffler, H., Dollfuss, H.: The benthic communities of Neusiedlersee (Austria). Verh.int.Verein. theor.angew.Limnol. 17, 201-208 (1969).

100 Schöttler, U.: Über den Anaerobiosestoffwechsel von Tubifex. Verh.Dtsch.Zool.Ges. 1974, 271-274 (1975).

101 Schöttler, U., Schroff, G.: Untersuchungen zum anaeroben Glykogenabbau bei Tubifex tubifex M. J.Comp.Physiol. B. 108, 243-254 (1976).

102 Seidman, I., Entner, N.: Oxidative enzymes and their role in phosphorylation in sarcosomes of adult Ascaris lumbricoides. J.Biol.Chem. 236, 915-919 (1961).

103 Simpson, J.W., Awapara, J.: Phosphoenolpyruvate carboxykinase activity in invertebrates. Comp.Biochem.Physiol. 12, 457-464 (1964).

104 Simpson, J.W., Awapara, J.: The pathway of glucose degradation in some invertebrates. Comp.Biochem.Physiol. 18, 537-548 (1966).

105 Smith, M.H.: Do intestinal parasites require oxygen? Nature, Lond. 223, 1129-1132 (1969).

106 Stokes, T.M., Awapara, J.: Alanine and succinate as end-products of glucose degradation in the clam Rangia cuneata. Comp.Biochem.Physiol. 25, 883-892 (1968).

107 Taylor, A.C.: Burrowing behaviour and anaerobiosis in the bivalve Artica papillosa (L.). J.Marine Biol.Assoc. 56, 95-109 (1976).

108 Theede,H., Ponat, A., Hiroki, K., Schlieper, C.: Studies on the resistance of marine bottom invertebrates to oxygen deficiency and hydrogen sulfide. Mar.Biol. 2, 325-337 (1969).

109 Thillard, G. van den, Kesbeke, F., Waarde, A. van: Influence of anoxia on the energy metabolism of goldfish Carassius auratus (L.). Comp.Biochem.Physiol. 55A, 329-336 (1976).

110 Utter, M.F., Kolenbrander, H.M.: Formation of oxalacetate by CO_2 fixation on phosphoenolpyruvate. The Enzymes Vol. 6, 117-168 (Boyer, P.D. (ed.), Academic Press New York, 3rd ed. (1972).

111 Wadsö, I.: Microcalorimeters. Quart.Rev.Biophys. 3, 383-427 (1970).

112 Ward, C.W., Schofield, P.J., Johnstone, I.L.: Carbon dioxide fixation in Haemonchus contortus larvae. Comp. Biochem.Physiol. 26, 537-544 (1968).

113 Watts, S.D.M., Fairbairn, D.: Anaerobic excretion of fermentation acids by Hymenolepis diminuta during development in the definitive host. J. Parasitol. 60, 621-625 (1974).

114 Wieser, W., Kanwisher, J.: Ecological and physiological studies on marine nematodes from a small salt marsh near Woods Hole, Massachusetts. Limnol.Oceanogr. 6, 262-270 (1961).

115 Wieser, W., Ott, J., Schiemer, F., Gnaiger, E.: An ecophysiological study of some meiofauna species inhabiting a sandy beach at Bermuda. Marine Biology 26, 235-248 (1974).

116 Wijsman, T.C.M.: pH fluctuations in Mytilus edulis L. in relation to shell movement under aerobic and anaerobic conditions. Proc.9th Europ.mar.biol.Symp., 139-149, Aberdeen Univ.Press (1975).

117 Wijsman, T.C.M.: Adenosine phosphates and energy charge in different tissues of Mytilus edulis L. under aerobic and anaerobic conditions. J.comp.Physiol. 107, 129-140 (1976).

118 Wijsman, Z.C.M., Zwaan, A.de, Ebberink, R.H.M.: Adenylate energy charge in Mytilus edulis L. during exposure to air. Biochem.Soc.Trans. 4, 442-443 (1976).

119 Wilson, M.A., Cascarano, J.: The energy-yealding oxidation of NADH by fumarate in submitochondrial particles of rat tissue. Biochem.Biophys.Acta 216, 54-62 (1970).

120 Wood, H.G.: Transcarboxylase. The Enzymes, Vol. 6, 83-115 Boyer, P.D. (ed.), Academic Press,New York,3rd ed. (1972).

121 Zebe, E.: In vivo-Untersuchungen über den Glucose-Abbau bei Arenicola marina (Annelida, Polychaeta). J.comp. Physiol. 101, 133-145 (1975).

122 Zee, D.S., Zinkham, W.H.: Malate dehydrogenase in Ascaris suum: Characterization, ontogeny, and genetic control. Arch.Biochem.Biophys. 126, 574-584 (1968).

123 Zoeten, L.W. de, Posthuma, D., Tipker, J.: Intermediary metabolism of the liver fluke, Fasciola hepatica. I. Biosynthesis of propionic acid. Hoppe-Seyler's Z. physiol.Chem. 350, 683-690 (1969).

124 Zoeten, L.W. de, Tipker, J.: Intermediary metabolism of the liver fluke Fasciola hepatica II. Hydrogen transport and phosphorylation. Hoppe Seyler's Z.physiol.Chem. 350, 691-695 (1969).

125 Zs.-Nagy, I.: The distribution of cytosomes in the non nervous tissues of Anodonta cygnea L. (Mollusca, Pelecypoda). Annal.Biol.Tihany 40, 121-134 (1973).

126 Zs.-Nagy, I.: Adenosine phosphate concentrations and carbohydrate consumption in the tissues of Anodonta cygnea L. (Mollusca, Pelecypoda) under normal and anoxic conditions. Acta Biochim.Biophys.Acad.Sci. Hung. 8, 143-151 (1973).

127 Zs.-Nagy, I.: Some quantitative aspects of oxygen consumption and anaerobic metabolism of molluscan tissues. A review. Comp.Biochem.Physiol. 49A, 399-405 (1974).

128 Zs.-Nagy, I., Ermini, M.: Oxidation of $NADH_2$ by the lipochrome pigment of the tissues of the bivalve Mytilus galloprovincialis (Mollusca, Pelecypoda). Comp.Biochem. Physiol. 43B, 39-46 (1972).

129 Zs.-Nagy, I., Ermini, M.: ATP production in the tissue of bivalve Mytilus galloprovincialis (Pelecypoda) under normal and anoxic conditions. Comp.Biochem.Physiol. 43B, 593-600 (1972).

130 Zwaan, A. de: Pyruvate kinase in muscle extracts of the sea mussel Mytilus edulis L. Comp.Biochem.Physiol. 42B, 7-14 (1972).

131 Zwaan, A. de, Bont, A.M.Th. de: Phosphoenolpyruvate carboxykinase from adductor muscle tissue of the sea mussel Mytilus edulis L. J.Comp.Physiol. B 96, 85-94 (1975).

132 Zwaan, A. de, Bont, A.M.Th. de, Kluytmans, J.H.F.M.: Metabolic adaptations on the aerobic-anaerobic transition in the sea mussel, Mytilus edulis L. Proc. 9th Europ.mar.biol.Symp., 121-138, Aberdeen Univ. Press (1975).

133 Zwaan, A. de, Holwerda, D.A.: The effect of phosphoenolpyruvate, fructose-1,6-diphosphate and pH on allosteric pyruvate kinase in muscle tissue of the bivalve Mytilus edulis L. Biochim.biophys.Acta 276, 430-433 (1973).

134 Zwaan, A. de, Holwerda, D.A., Addink, A.D.F.: The influence of divalent cations on allosteric behaviour of muscle pyruvate kinase from the sea mussel Mytilus edulis L. Comp.Biochem.Physiol. 52B, 469-472 (1975).

135 Zwaan, A. de, Kluytmans, J.H.F.M., Zandee, D.I.: Facultative anaerobiosis in molluscs. Biochem.Soc.Symp. 41, 133-168 (1976).

136 Zwaan, A. de, Marrewijk, W.J.A.: Anaerobic glucose degradation in the sea mussel Mytilus edulis L. Comp.Biochem. Physiol. 44B, 429-439 (1973).

137 Zwaan, A. de, Wijsman, T.C.M.: Anaerobic metabolism in bivalvia (Mollusca). Characteristics of anaerobic metabolism. Comp.Biochem.Physiol. 54B, 313-324 (1976).

138 Zwaan, A. de, Zandee, D.I.: Body distribution and seasonal changes in the glycogen content of the common sea mussel Mytilus edulis. Comp.Biochem.Physiol. 43A, 53-58 (1972).

139 Zwaan, A.de, Zandee, D.I.: The utilization of glycogen and accumulation of some intermediates during anaerobiosis in *Mytilus edulis* L. Comp.Biochem.Physiol. 43B, 47-54 (1972).

6.2 Microcalorimetric Investigations of Aquatic Biotopes

F. Tiefenbrunner

INTRODUCTION

The largest fraction of nutrients dissolved in natural waters are organic substances. Dissolved organic carbon (DOC) may be produced within the system by decomposition of organic detritus or by excretion from members of the biotopes. It has been shown (1), that algae excrete as much as 80% of their photosynthetic products, macrophytes in the litoral zone of a lake may also excrete a large percentage of their photosynthetic products. It has been reported (2), that zooplancton is also a significant source of soluble organic carbon. In addition to production within the system (autochthonous)there is introduction from precipitation and run - off water containing plant material and waste products introduced by man.

Decomposition, transformation and mineralization of organic material in the aquatic environment is performed by heterotrophic micro - organisms, mainly bacteria, which are the main agents for return of dead organic matter to the soluble state (3). However bacteria can directly utilize DOC as nutrients returning DOC to the biological system.

The specific requirement of measuring heterotrophic activity is only partly possible using direct methods. Besides the measurement of the biochemical oxygen demand (BOD) and enzymkinetic studies of uptake of labelled organic substrates(4), heterotrophic activity is calculated mainly from biomass estimations in combination with chemical parameters (5). Microcalorimetry is a nonspecific method for the direct measurement of biological activity, and therefore could offer advantages for examining the heterotrophic activity in natural waters and water treatment plants. Previous experiments comparing calorimetric measurements on biotopes, including BOD, chemical oxygen demand (COD), enzyme kinetic uptake of glucose and manometric methods showed that commercially available calorimeters have a similar sensitivity as manometric measurements.

Calorimetry at the present time is restricted to measurement of eutrophic biotopes having plate counts (Agar, 22°C) in the order of $1o^5$ cells /ml. However calorimetry is a useful tool for controlling sewage processing, effluent monitoring of purified sewage water and following heterotrophic activity during algal blooms. Using an indirect method it is also possible to estimate toxic compounds in water.

EXPERIMENTAL

Calorimeters

A flow calorimeter (LKB,Bromma, Typ 21o7) was operated at 3o ml/h, using a steel flow cell of 1 ml volume. Three ampule calorimeters (6), 7 ml volume, were simultaneously used in some experiments, and another (LKB,Bromma,prototype) of 4 ml volume for other experiments. The calorimeters had a heat effect sensitivity better than o,o5 µV / µW. In operating the ampule calorimeters the time between loading and measurement was 5o minutes.

Investigation Areas

Two different types of sewage treatment plants were examined. The first plant (A) was about 12o ooo inhabitation equivalents, with biological treatment by medium loaded trickling filters, chemical treatment with aluminium sulphate followed by passage through three oxidation ponds. The effluent drained into a river flowing at 2oo - 4oo l/sec. Samples were taken at the outflow from the primary sedimentation basin (" after prim. sed."), at the outflow from the secondary sedimentation basin (" after sec. sedim."), the third oxidation pond and from the river upstream (" river") and downstream from the effluent outlet.

The second plant (B) was of 34o ooo inhabitation equivalents operated with biological treatment using the activated sludge process. Simultaneous flocculation by adding ferrous sulphate in the grit chambers was used for reduction of the phosphorus content of the sewage. The effluent after passage through a lake system entered the Baltic Sea. Samples were taken at the outflow of the primary and secondary sedimentation basins and from lake water located at about 25 m from the effluent pipe. All samples for examination were stored in plastic bottles for about 2.5 h between sampling and processing. The calori - metric experiments were made in several series. A comparison was made between ampule and flow calorimeters using samples from the secondary sedimentation basin (after biological treatment) and the effluent from the oxidation pond to the river in plant (A). The effect of glucose addition was examined using samples from the primary sedimentation basin (after mechanical purification), the secondary sedimentation basin and outlet from the oxidation pond to the river in plant (A). Escherichia and Pseudomonas species isolated from the oxidation pond (Nr. 3) in plant (A)were prepared as suspensions in physiological saline and the effect of glucose additions (1mg/ml) were measured in the flow calorimeter. The effect of temperature (25°C and 32°C) on activity to glucose was measured on samples from the primary sedimentation basin and the secondary sedimentation basin in plant (B). Similar samples were examined in relation to glucose degration and temperature were sampling time was varied.

RESULTS

In figure 1), the thermograms are shown from flow- and ampule calorimeters using samples taken from the effluent from the secondary sedimentation basin and 3rd oxidation pond (plant A). The flow thermogram gave a smaller signal in each case due to the smaller volume measured. However, the bacterial density was high enough for measurable heat effects. Samples from the outlet from the oxidation pond did not show a high enough heat effect in the flow calorimeter to make any conclu sions during the first six hours of incubation. The results comparing samples with and without added Glucose are shown in figure 2 a). Effluent from the primary sedimentation basin (after mechanical purification) and effluent from the secondary sedimentation basin (after biological purification) from plant (A) were used and measured in ampule calorimeters at 25° C . Both samples showed an increase in heat production after the addition of glucose. A sample from the outflow of

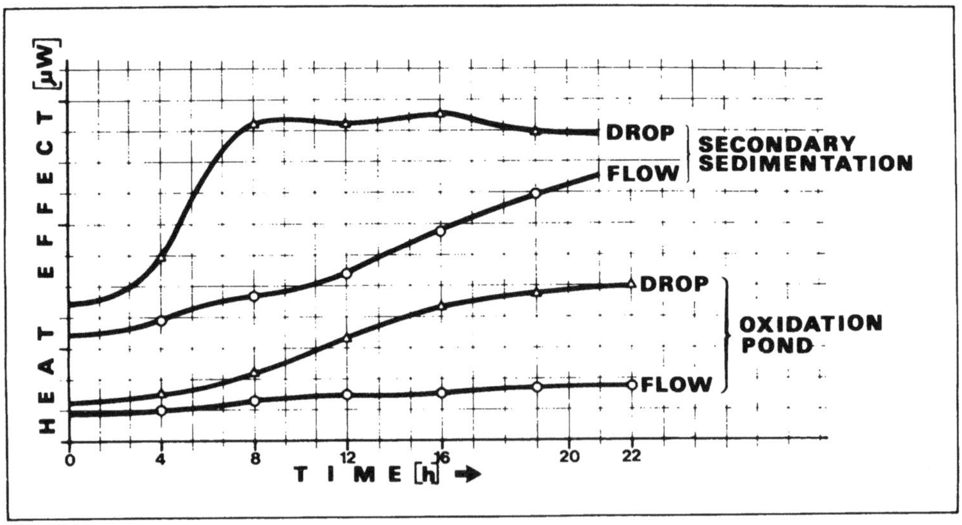

Figure 1) Comparison of thermograms from flow and ampule cal.

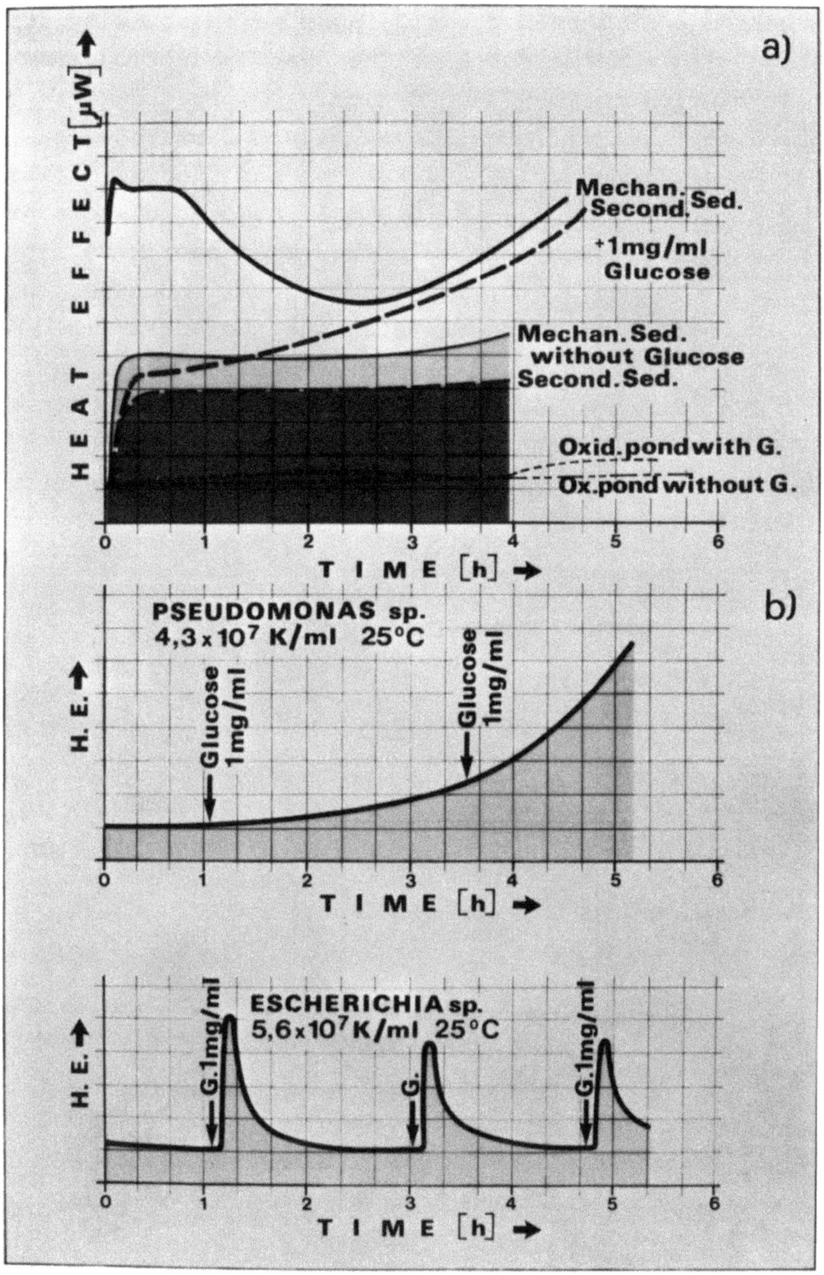

Figure 2a) Comparison of samples with and without added gluc.
Figure 2b) Difference in heat response of two bacteria species

the oxidation pond showed a small increase in activity after incubation with glucose for 6 h. The heat production from the different samples could not be directly compared because they were taken on three different days. Figure 2 b) shows the thermograms of Escherichia- and Pseudomonas species isolated from the first oxidation pond (plant A)suspended and stored for 4 h in physiological saline before examination by flow - calorimetry. Glucose (1mg/ml) was added as indicated on the figure. The effect of temperature on the thermograms of samples from the primary and secondary sedimentation basins (plant B) to which glucose had been added are shown in figure 3). The heat output at 32° C is significantly higher than at 25° C , the maximum activity of the sample being obtained within 6 h at 32° C.

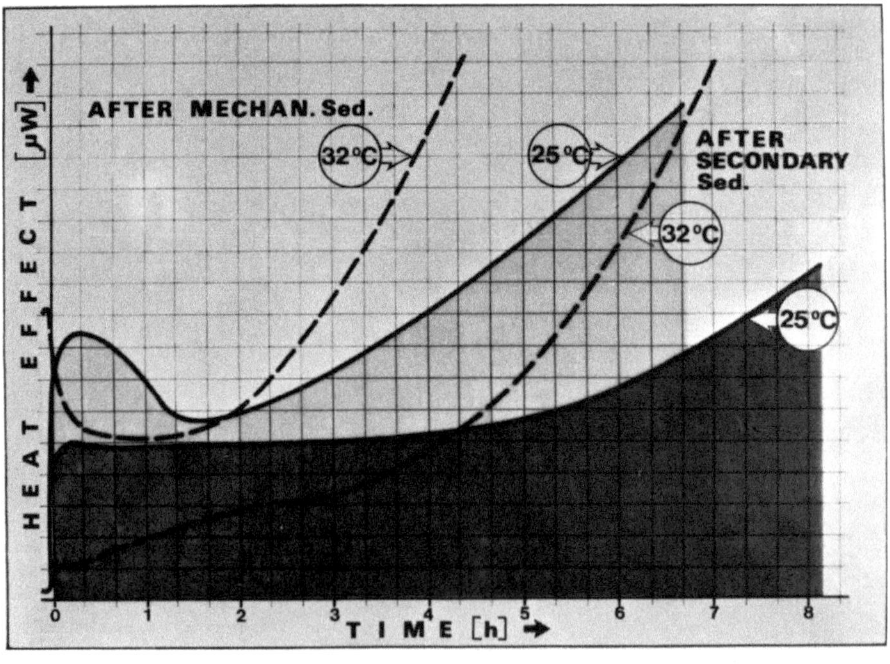

Figure 3) The effect of temperature on the thermograms of samples from primary and secondary sedimentation basins.

Figure 4) shows activity of biologically treated sewage water from plant (A), sampled on sunday morning and monday afternoon, incubated with glucose at 25° C in an ampule calorimeter

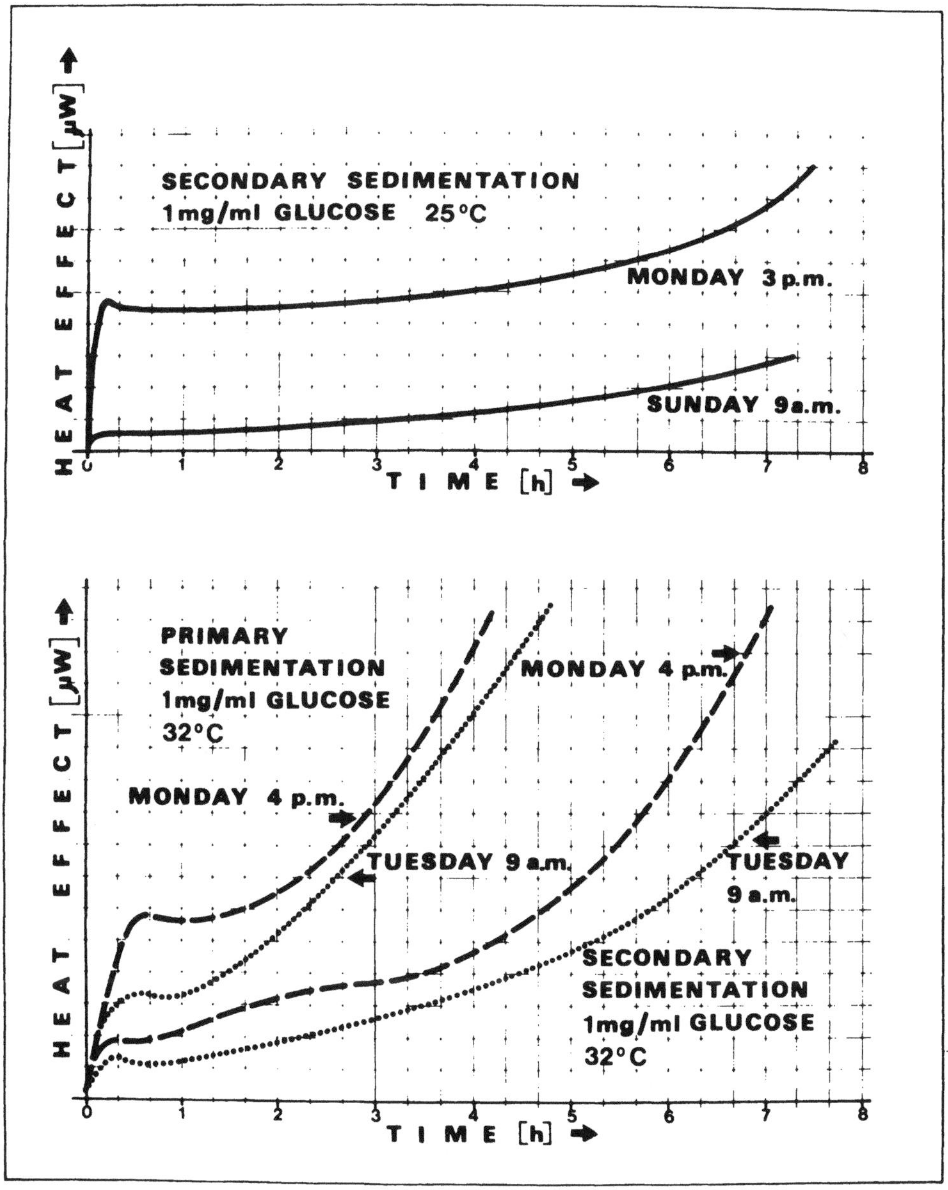

Fig. 4) Heat production of samples taken at 3 different days.

The first thermogram shows a very small response compared to the second sample reflecting the different rate of influx of raw sewage to the plant. Samples from the primary sedimentation basin and the secondary sedimentation basin from plant (B) incubated with glucose at 32° C are shown in the lower part of figure 4). The samples were taken on consecutive week days, monday afternoon and tuesday morning, showing smaller differences in response than between sunday and monday samples. Table 1) combines heat effects, COD ,BOD and PO_4-P contents at different stages of sewage purification.

	AFTER PRIMARY SEDIM.			AFTER SECONDARY SEDIM.			OXYDATION POND			RIVER		
	3^h	6^h	8^h	3^h	6^h	8^h	3^h	6^h	8^h	3^h	6^h	8^h
HEAT EFFECT in μW GLUCOSE 1mg/ml 3ml SAMPLE,25°C	55,2	111,3	159,3	37,3	50,4	79,6	8,1	7,6	10,5	5,3	6,7	7,0
HEAT EFFECT in μW GLUCOSE 1mg/ml 3ml SAMPLE,32°C	84,6	244,4	-	25,3	85,5	211,5	-	-	-	-	-	-
COD mg/L	270			135			28 (15,2)			8,4		
BOD mg/L	120			36			9,8			4,1		
PO_4- P mg/L	-			-			0,38			0,29		

Table 1

CONCLUSIONS

Calorimetry was first applied to the study of aquatic biotopes several years ago, when it was hoped to measure heterotrophic activity in many different aquatic systems. However, after many investigations the interest in application of calorimetry to this field decreased, mainly because of limits in detection with commercial calorimeters presently available. The instruments are restricted to use with highly activ biotopes exhibiting rapid substrate turnover. At biotope tempera tures (8 to 2o^{o} C)and without addition of a substrate that can be rapidly degraded, bacterial counts of 1o^{7}/ ml would be necessary to give a reproducible heat effect in a flow calorimeter (LKB,21o7).With glucose added to natural water samples cell densities of 1o^{5}/ ml are required. Samples suitable for calorimetric studies are therefore only available from lakes during algal blooms (4) or biotope areas close to the influ ence of sewage waters. Samples from other aquatic biotopes require up to 1oo times concentration to obtain satisfactory heat effects for calorimetric measurement. Concentrations of this magnitude could result in large disturbances to the system reducing confidence in the meaning of calorimetric measurements. Comparisons between flow calorimeters and manometric techniques showed them to be of the same order of sensitivity. The sensitivity of manometric techniques can be increased to a limited extend by increasing easily the size of the reaction vessel. Automatic operation and recording of results from manometric respiration equipment requires much handling and supervision and the auxillary equipment required increases the cost to about the same as of commercially available microcalorimeters. A flow calorimeter however, is easily adapted to automatic control requiring less supervision and simplicity of operation. From this point of view continuous control by flow calorimetry allows optimal monitoring of f.e. sewage disinfection and water purification in potable water plants from reservoir or river.

The development of calorimeters towards the requirements of limnologists and microbiologists dealing with water would be of importance to a variety of applications, including nitrification, denitrification and metabolism in the H_2S - zone. Increasing measurement volumes and using multichannel instruments would allow simultaneous screening of the effect of substrates, different oxygen tensions and different sample concentrations. Such instruments would be useful for studying anaerobic and aerobic sludge fermentation and as a warning system for poisons in sewage water.

However, the main requirement is for a fast responding system for measuring biological activity in eutrophic surface waters adding productivity parameters to chemical analysis and biomass estimations.

REFERENCES

(1) Fogg, G. E.: The role of algae in organic production in aquatic environments. Brit. Phycological Bulletin 2, 195 - 2o5 (1963).

(2) Webb, K. L., Johannes, E. E.: Studies of the release of dissolved free amino acids by marine zooplankton. Limnology and Oceanography 12, 376 - 382 (1967).

(3) McCoy, E.: Bacteria in lakes: Populations and functional relations, p. 311 - 339, in : Causes, consequences correctives. National Academy of Sciences,Washington, D.C. (1969).

(4) Tiefenbrunner, F.,Düsing, F.,Krambeck, H. J. und Overbeck, J.: The use of an enzyme kinetic method for bacteriological examinations of heavy polluted surface waters. Zbl.Bakt.Hyg., I.Abt. Orig. 161,498-518 (1976).

(5) Tiefenbrunner, F. und Rott, E.: Biological problems of a manmade small lake. Zbl.Bakt.Hyg.,I.Abt.Orig. 16o, 268 - 285 (1975).

(6) Wadsö, I.: A microcalorimeter for biological analyses. Science Tools 21, 18 - 21 (1974).

6.3 Combustion Heat in Ecological Energetics. What Sort of Information can be Obtained?

W. d'Oleire-Oltmanns

Introduction

The goal of ecological energetics is to describe all organisms in terms of a single unit. Mapping energy flows through an eco-system is a way to reduce the complexity of the system to a manageable level. This approach, however, has been criticized on several points. Boyd and Goodyear (1), for instance, showed that aquatic plants with nearly the same caloric value differ greatly in their chemical composition. The protein content, for example, ranged from 4 to 21 %. Verduin (2) has also pointed out that the combustion heat of plant material includes energy from compounds such as cellulose and lignin, which cannot be utilized by most herbivores. In short, the heat of combustion yields little information about the nutritive value of the mixed-substance organism.

The maximal heat of combustion for a mixed-substance organism is about 9500 cal/g (lipids). The minimum is determined by the ash content, with 2000 cal/g (glycine) as the minimum for ash-free data. However, as all organisms are composed of lipids, proteins und sugars, the ash-free heats of combustion can actually vary only within a narrow range. The large variation in plant caloric values uncorrected for ash content can be ascribed to low energy oxalate salts (3). The purpose of this work is to determine the importance of these factors in the results of ecological energetics.

Methods

The heats of combustion determined for various groups of organisms have been compiled to determine the distribution of the values. All data except those for Alnus glutinosa and Agelastica alni have been taken from Cummins and Wuycheck (4). The data were divided into 10 % classes for the purpose of frequency analysis. A logarithmic scale was used for the abscissa to reflect the scaling of trophic levels. The difference between 400 and 800 cal/unit, for instance, is very important to the "predator" feeding at that level, while the difference between 4400 and 4800 cal/unit is much less so to the predator feeding at the higher level. On the linear scale, the low-energy data would be concentrated, while the high energy points would be spread out. The mean of the mean values given by Cummins and Wuycheck (4) for individual species was determined, and the individual means falling within ± 10 % of the composite mean were counted.

The rationale for choosing the 10 % classes and the range is that 1) the microbombs used in ecological energetics have an error of about ± 5 %, and 2) the errors in every operation from sampling to burning add to the uncertainty. 10 % thus seems a reasonable estimate. There will also be a contribution to the uncertainty from the variability of the biological material. To determine energy flow, the values for an average representative of a population are needed. Systematic errors in sampling the population thus increase the uncertainty.

Results and Discussion

Plant material

Fig. 1 summarizes the combustion data for living plant material obtained from Cummins and Wuycheck (4). All data which are not from whole plants or vegetative parts have been eliminated. Seeds and reproductive material were eliminated because they contain large amounts of storage substances. They also represent only a small amount of the plant biomass and are in most cases only seasonally present. The non-ash free data falls into a rather broad distribution, due to the variation in ash contents, as will be shown later.

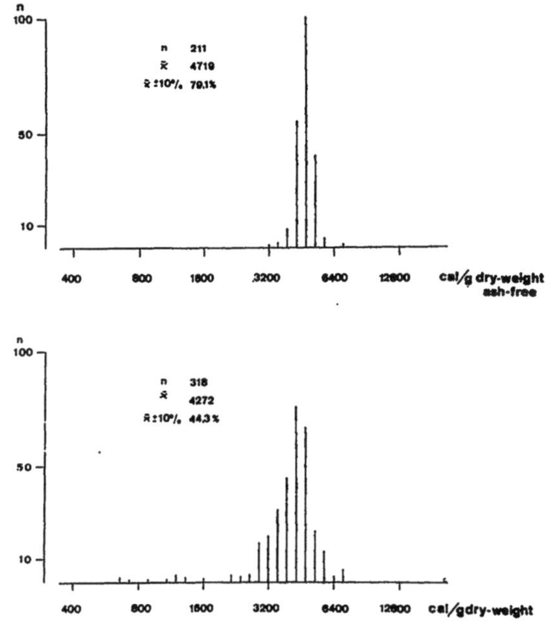

Fig. 1
Living plant material. Frequency distribution of the caloric values derived by combustion. Data from Cummins and Wuycheck (4). The columns indicate the middle of a class, ·x indicates the mean, x ± 10 %: The amount of data which have been found within this range.
Abscissa: logarithm scale
Ordinate: number of data in the concerning class

For a more differentiated view, the data have been divided into two classes: aquatic plant material (bacteria and cryptogame plants, fig. 2) and phanerogame terrestial plants (fig. 4). Most of the low-energy, non-ash-free data from fig. 1 comes from the aquatic plants. A plot of ash content versus dry weight (fig. 3) shows a strong inverse correlation between ash content and heat of combustion. The mean values for terrestial plants are higher than for aquatic plants, and the non-ash-free data is more concentrated. Very high ash contents are not observed here.

Since most of the data so far collected falls within reasonably narrow limits, one can predict that values determined in the future will also fall within the same limits, even though exceptional values have been found (5, 6). In the future, a test burning should suffice.

An example of the relative constancy of the heats of combustion for a single species, I have measured the values for elder (Alnus glutinosa) over the whole vegetation period of 1974. The distribution of the data is

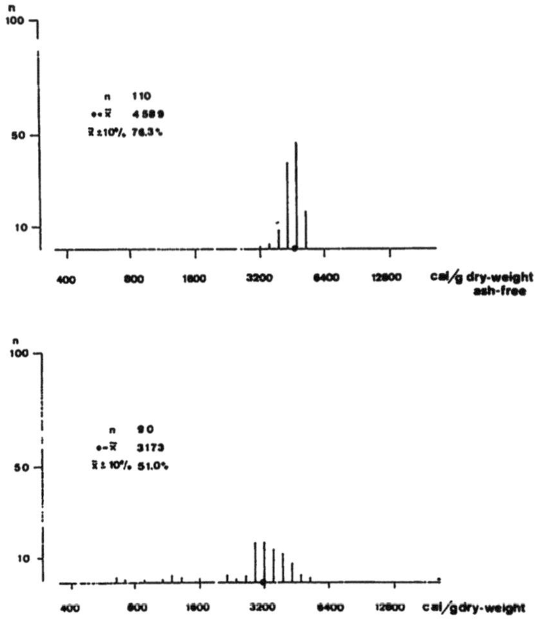

Fig. 2
Living aquatic material from bacteria and cryptogame plants (See fig. 1)

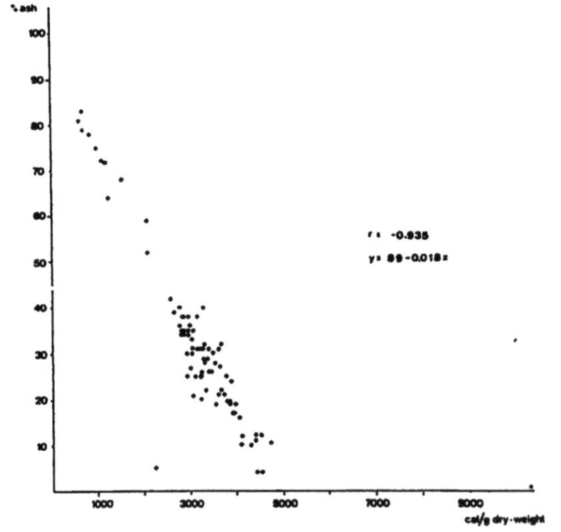

Fig. 3
Correlation between ash content and cal/g dry weight for the data from fig. 2
Abscissa: calories per gram dry weight
Ordinate: ash content

shown in fig. 5. The mean, which is derived from individual measurements, is nearly the same as for phanerogame plants in general. 95 % of the points fall within the $\pm$ 10 % range. My prediction is that vegetative material from phanerogames will nearly always fall within or near this range. For cryptogame aquatic material this is not as probable, but here, too, a random measurement will give as accurate data as necessary.

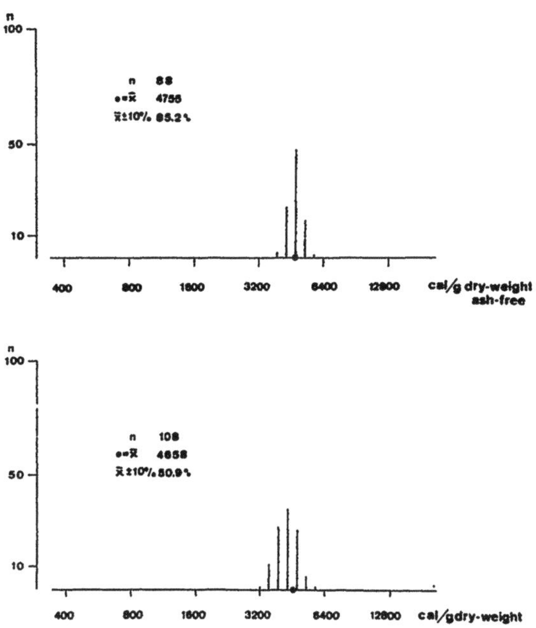

Fig. 4
Living terrestrial phanerogame plant material (See fig. 1)

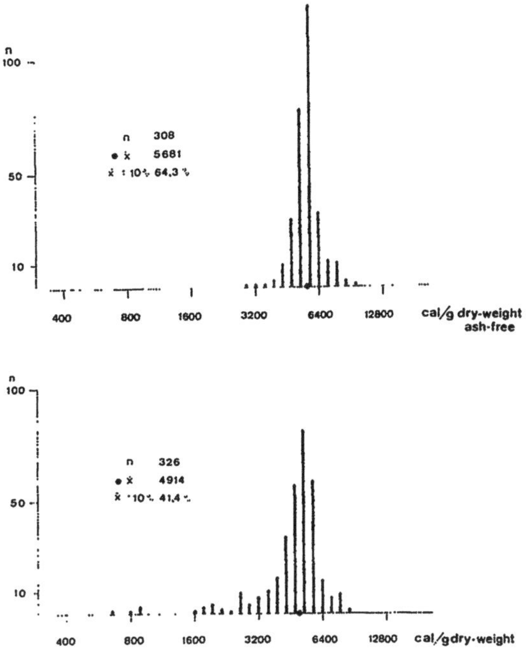

Fig. 6
Entire animals. Data from protozoans to mammals (See fig. 1)

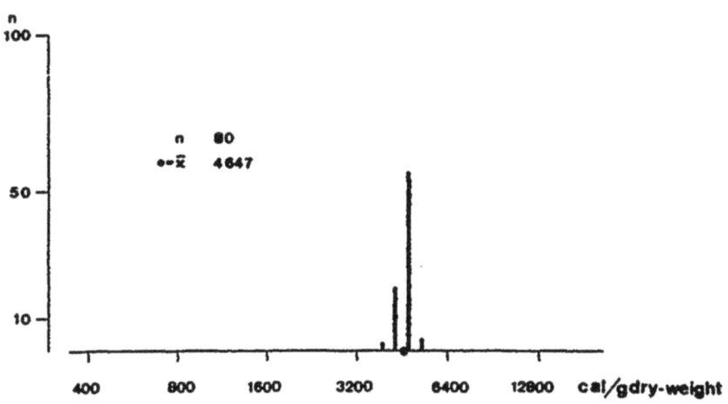

Fig. 5
Frequency distribution of caloric values from leaves of elder (Alnus glutinosa) collected during the whole vegetation period 1974 (Coordinates see fig. 1)

Animal material

Unlike plants, animals incorporate highly diverse structural elements, such as endo- and exoskeletons, chitin or calcium shields, none of which occur in plants. The distribution of all available data from entire animals, from protozoans to mammals, is shown in fig. 6. We find a broader distribution than for plant material (fig. 1), as might be expected from the greater structural diversity. Here, too, we would like to subdivide the data for greater differentiation, but sufficient data for this purpose are available only for arthropods (fig. 7). The spread of the data is large, due to the wide variation in the crustacean data (fig. 8). I believe this scatter in the crustacean data is due to methodological difficulties with tissues in which high ash contents can be found (7).

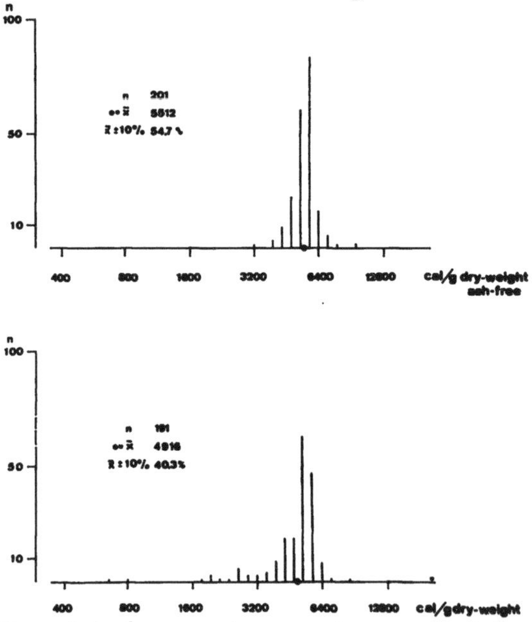

Fig. 7
Arthropod data (all stages, entire animals only) (See fig. 1)

When data for crustaceans is removed from the arthropod data, the distribution shown in fig. 9 is observed. The data is still representative of widely varying arthropods.

Over the vegetation period 1974 I measured larvae and adults from a chrysomelid beetle (Agelastica alni) feeding on elder (fig. 10). For this

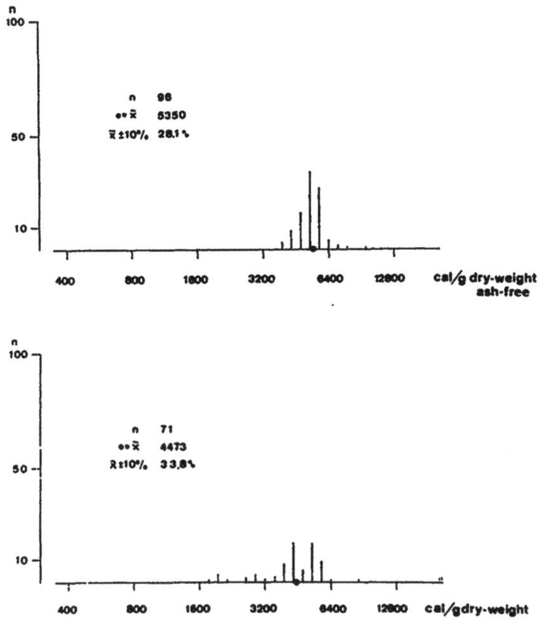

Fig. 8
Crustacean material (all stages, entire animals only) (See fig. 1)

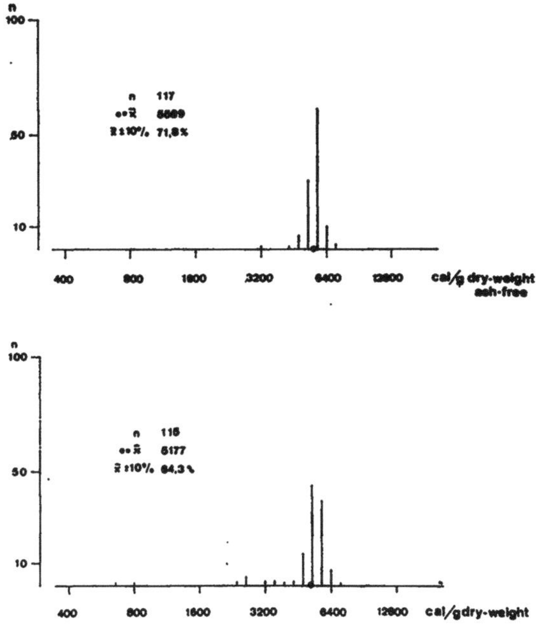

Fig. 9
Athropod material apart fram crustacean material (fig. 9 = fig. 7 - fig. 8) (See fig. 1)

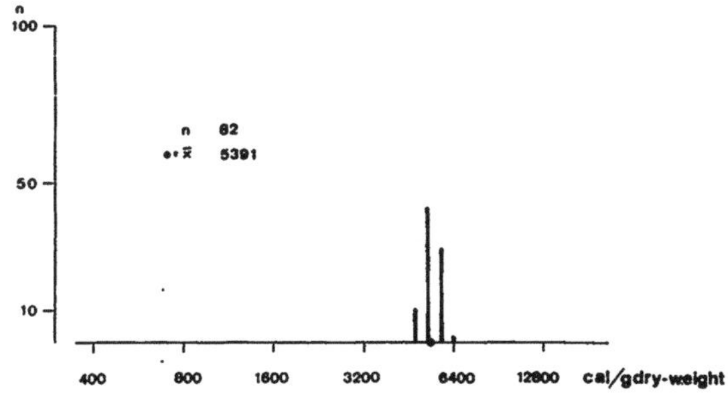

Fig. 10
Frequency distribution of the caloric values from larvae and adult Agelastica alni (Chrysomelidae) collected during the vegetation period 1974 (Coordinates see fig. 1)

beetle we find 72 % of all points within 10 % of the arthropod mean calculated without the crustaceans. The mean of the beetle data differs from the arthropod mean by only 3 %.

Conclusion

What is the justification for using generalized caloric values for organisms in ecological energetics? Basically, it is that the other errors involved in determining an energy budget or the energy flow for a population are likely to outweigh by far the error incurred by taking a generalized caloric value. First, in order to determine A = P + R (where A is assimilation in the ecological sense, P is production and R is respiration), we need at least the respiration data, which must be obtained in the laboratory. There we must choose a working temperature which is something like a mean temperature from the field. This creates at least an uncertainty, if not an error.

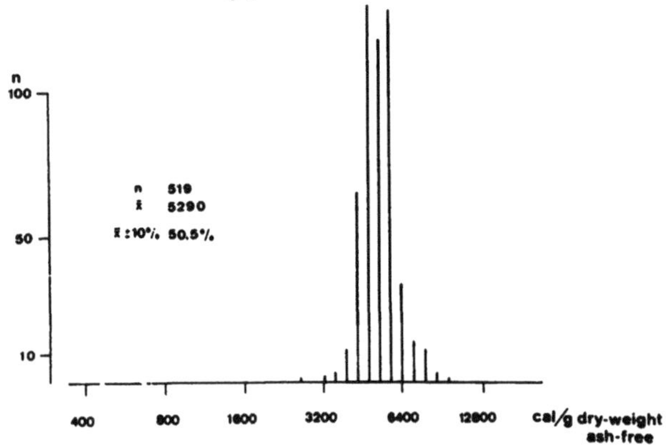

Fig. 11
Frequency distribution of all ash-free data shown in the figures above (From bacteria to mammals) (See fig. 1)

Second, for ecosystem research we need the population density, or better still, the fluctuations in the population density. Determination of the population makes the decisive contribution to the entire error of such an ecological study, since uncertainties of 50 % or more in the population level are often reported. Thus it seems that the relatively high concentration of caloric data in the range of the mean ± 10 % justify working with the mean values (8). This is supported by the consideration that we have to base our figures on the "average individual", and by the following characteristics of the data:

1. The heats of combustion of entire organisms seem to be similar if the ash-free data are used (fig. 11). (The figure shows data derived by the same criteria as mentioned above, from bacteria to mammals).
2. The means for plant and animal material are rather different.
3. The data are levelled because of mostly small random samples and the uncertainty of the method.
4. Sugars and proteins, with similar caloric contents, make up the greater part of organisms. The differences in calories within a systematic group are due mainly to the different lipid and ash contents and/or in the interference of salts in the combustion (3, 7).

Given the difficulties in population determination and the limited information provided by the heat of combustion for the mixed-substance organism, it follows that with this approach we obtain at best very limited information about connections in an ecosystem. Therefore, it is my opinion that if we use a simplified approach to these complex systems, we should not place undue value upon greater accuracy, as is often attempted for combustion heats used in ecological energetics. Further improvement in calorimetric accuracy would be a step towards completeness of measurement, but would do nothing to provide deeper understanding of ecosystems.

Acknowledgement

I want to thank the members of the Institute of Biophysics, FU Berlin, especially Mr. Lamprecht, who made this work possible. Mr. Haarhaus helped me to a great extent by discussing the problems with me.

Summary

The caloric contents of organisms fall within a range limited on the low side by glycine (2000 cal/g) and on the high side by fats (9500 cal/g). These extremes are rarely measured, because organic tissue is usually composed of a mixture of sugars, proteins and fats. The caloric values for mixed-substance organisms thus fall within a reduced range. Frequency analyses show that for certain systematic groups, the caloric values cluster about the mean. Because the data is levelled, we must question the value of highly accurate calorimetric measurements as a tool for ecological energetics.

References

1. Boyd, C. E., Goodyear, C. P.: Nutritive quality of food in ecological systems
Arch. Hydrobiol. 69, 256 - 270 (1971)

2. Verduin, J.: Caloric content and available energy in plant matter
Ecology 53, 982 (1972)

3. Darling, M. S.: Interpretation of global differences in plant caloric values
Oecologia 23, 127 - 139 (1976)

4. Cummins, K. W., Wuycheck, J. C.: Caloric equivalents for investigations in ecological energetics
Mitt. Internat. Verein. Limnol. 18, 1 - 158 (1971)

5. Bliss, L. C.: Caloric and lipid content in alpine tundra plants
Ecology 43, 753 - 757 (1962)

6. Larcher, W., Schmidt, L., Tschager, A.: Starke Fettspeicherung und hoher Kaloriengehalt bei Loiseleuria procumbens (L.) Desv.
Oecol. Plant, 8, 377 - 383 (1973)

7. Paine, R. T.: The measurement and application of the calorie to ecological problems
Ann. Rev. Ecol. System. 2, 145 - 164 (1971)

8. Runge, M.: Energieumsätze in den Biozönosen terrestrischer Ökosysteme
Scripta Geobotanica 4, 1 - 77 (1973)

6.4 Energy Flow and Efficiency Differences in Plants and Plant Communities

H. Lieth

INTRODUCTION

Plant growth and dry-matter production provide the basic energy for all living organisms on earth. The energy is stored in chemical substances that constitute the plant body. While the primary produced material is normally sugar, the bulk of the plant body is composed of a variety of compounds. These compounds differ widely in their combustion value. It is therefore advisable for energetic considerations in ecology and physiology to measure the energy content of produced materials in a time series along with dry-weight and chemical composition.

The following paper demonstrates the usefulness of such energy assessments at three different levels: 1) On the level of global biome productivity patterns; 2) On the level of energy depreciation during decomposition in different ecosystems; and 3) On the level of plant growth by comparing dry-weight and energy-sum development of a single plant. The data presented here were either obtained in the laboratory of the author by students or were obtained from the literature and comparatively evaluated by the author and his associates. Caloric values used in this paper were measured equal to or similar to the procedure described by Lieth (1).

THE GLOBAL NET ENERGY PRODUCTIVITY PATTERN

Net primary productivity can be defined in three ways: net photosynthesis (fixation of CO_2), net dry-matter productivity (dry-weight accumulation), and net energy accumulation (combustion value of the produced dry-matter). All three categories, related to the same surface area and time unit, yield one type of valuable productivity estimate. All three are related to each other but are not identical. Total net photosynthesis does not take into account that dry-matter contains ash and that different organic matter varies greatly in combustion value. The majority of productivity data is available as dry-matter weight, which is also preceding any energy determination. The increased amount of labor needed for calorimetry prevents most researchers from measuring energy values. The majority of energy estimates are therefore done by approximate conversions. There is, however, a large volume of combustion values available. The most complete compilation was recently published by Cummins and Wuycheck (2).

In 1964 the author published a primary productivity map of the world (3). This map was intended also to show the energy fixation pattern on earth. There were, however, not enough details available to the author at that time to convert dry-weight assessments into energy values. The map was constructed basically by converting classical vegetation maps into a productivity map using a set of productivity averages paired with vegetation units. In a similar approach we paired vegetation types with average energy fixation rates (4, 5, 6). The land vegetation types and their energy fixation rates are shown in Table I. This tabulation can be used to calculate the total energy fixation of the land areas, which amounts to 426.1×10^{18} cal per year.

The energy fixation rates for the individual vegetation categories in Table I can be used to convert a global vegetation map into a map of energy fixation rates. Although maps so constructed yield only crude generalized patterns, they have been proven useful in several other productivity related world maps.

In order to prepare a world map of energy fixation levels on land we used the computer world map developed earlier in our laboratory (4, 5). The map has been used to demonstrate the distribution of vegetation types listed in Table I (7). This map was recently converted by E. Box into an annual energy fixation map by using about 1200 stations evenly scattered over the continents for which the vegetation types were determined. The conversion of vegetation types into energy values was performed by a computer program called ECOSIEVE (Box, unpublished) using a slightly more detailed tabulation of energy fixation rates than shown in column 3 of Table I. The numbers so gained were then used as the E-values package of

TABLE I: Net energy fixation of the terrestrial vegetation of the world (after Lieth 1975), (6).

(1) Vegetation unit	(2) Area 10^6 Km^2		(3) Energy fixation rate 10^6 cal/m^2/y	(4) Total energy fixed per area and year 10^{18} cal	
Forest	50.0			277.0	
Tropical rain forest		17.0	8.2		139.4
Raingreen forest		7.5	6.3		47.2
Summergreen forest		7.0	4.6		32.2
Chaparral		1.5	3.9		5.9
Warm temperate mixed forest		5.0	4.7		23.5
Boreal forest		12.0	2.4		28.8
Woodland	7.0		2.8	19.6	
Dwarf and open scrub	26.0			10.2	
Tundra		8.0	0.6		4.8
Desert scrub		18.0	0.3		5.4
Grassland	24.0			60.0	
Tropical grassland		15.0	2.8		42.0
Temperate grassland		9.0	2.0		18.0
Desert (extreme)	24.0			0.1	
Dry desert		8.5	-		0.1
Ice desert		15.5	-		-
Cultivated land	14.0		2.7	37.8	
Freshwater	4.0			21.4	
Swamp and marsh		2.0	8.4		16.8
Lake and stream		2.0	2.3		4.6
Total for land	149.0			426.1	

the computer program SYMAP. The procedure of the preparation of our maps and their evaluation is discussed in Box (8).* The resulting map of annual energy fixation rates is shown in Figure 1. According to our custom to name the map after the conference town where the map was prepared for, this map is labeled as Berlin 1976 map.

*Detailed information will soon be available in the Ph.D. Thesis by E. Box at the University of North Carolina.

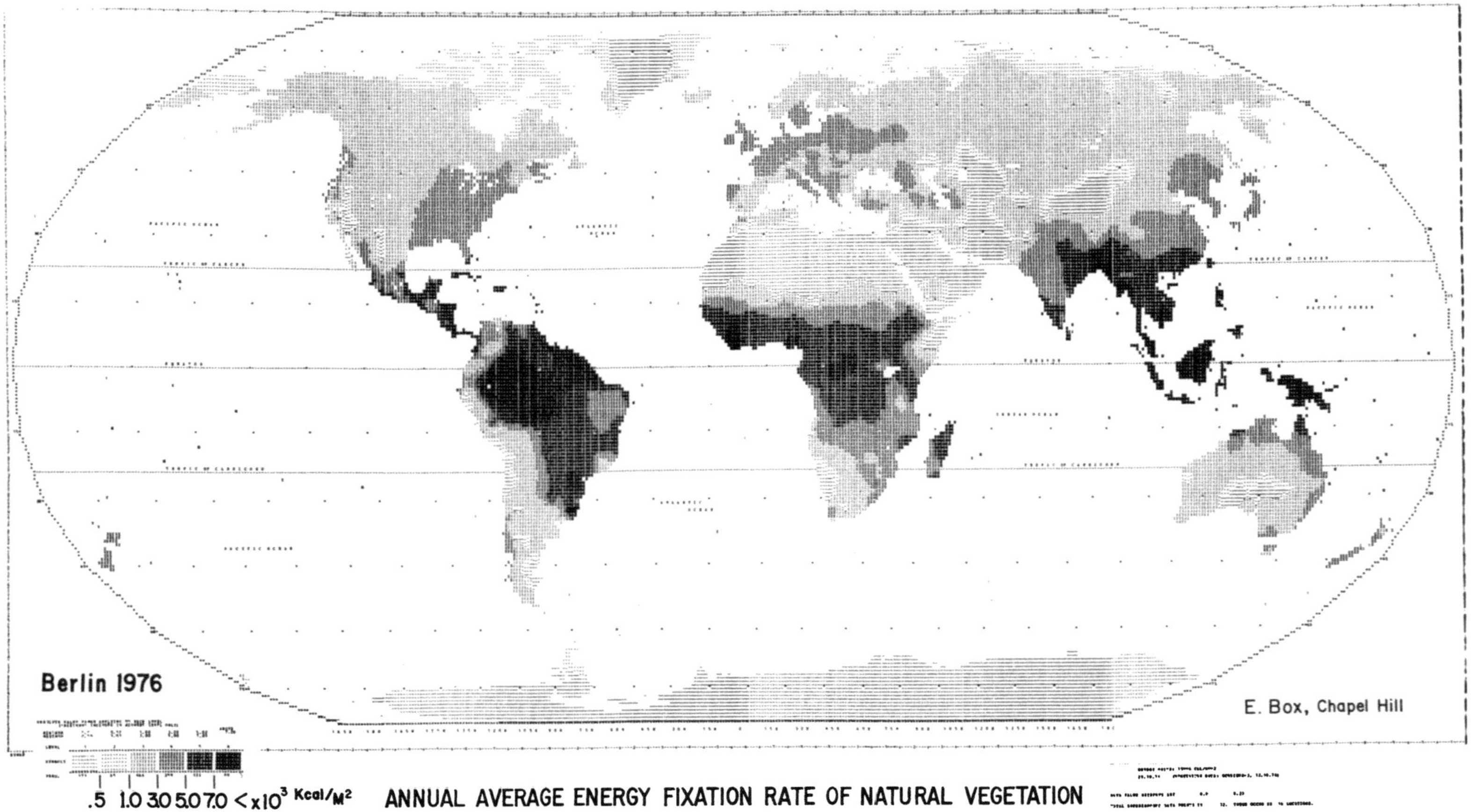

ANNUAL AVERAGE ENERGY FIXATION RATE OF NATURAL VEGETATION

Fig 1

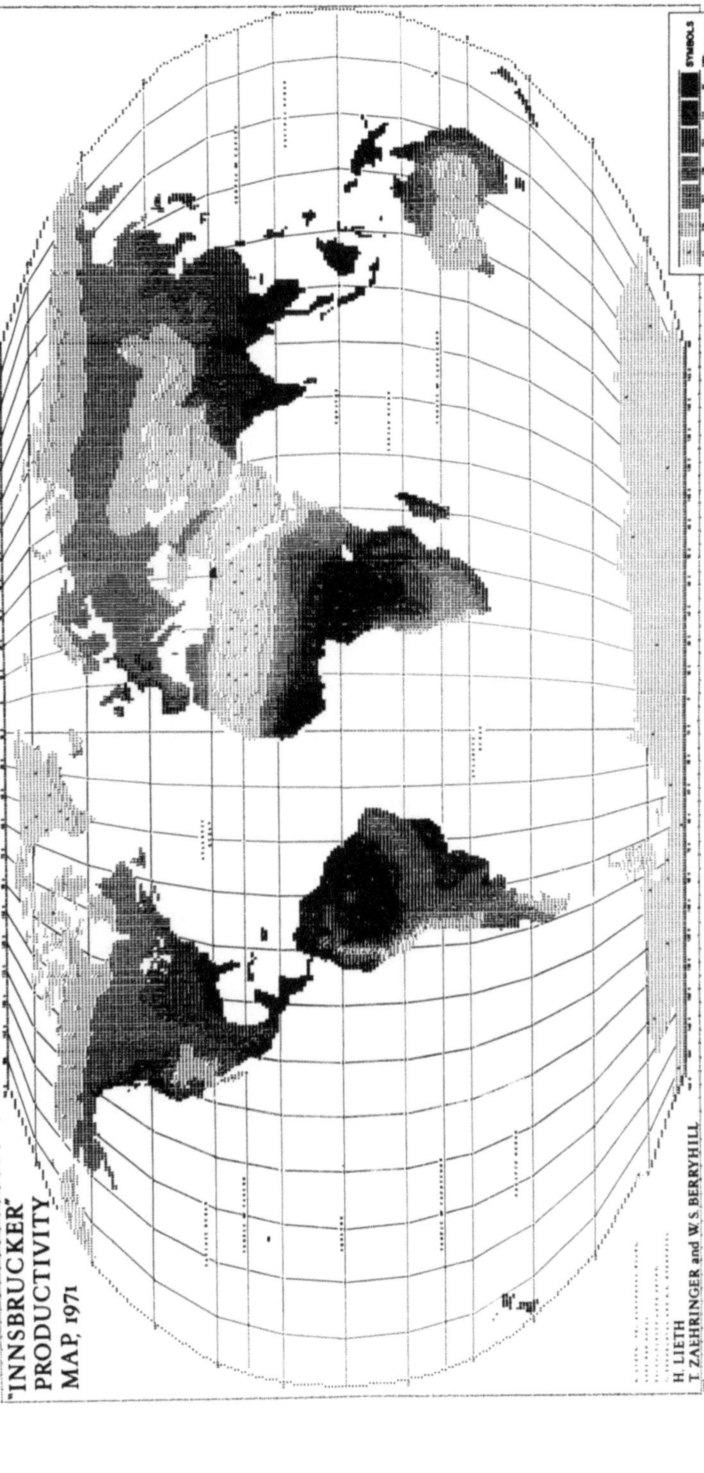
"INNSBRUCKER" PRODUCTIVITY MAP, 1971
H. LIETH
T. ZAEHRINGER and W.S. BERRYHILL
SYMBOLS

Fig 2

A quick comparison of the Berlin map (energy pattern) with the Innsbruck map, the computer simulation of our 1964 productivity map (4), (dry-weight productivity pattern), Figure 2, shows general similarities. The patterns are sensitive to the different numbers of intervals, the number of stations, and their positions used. One significant difference appears to be the narrower range of the tropical high-productivity belt versus the temperate zone forest belt. This reflects the fact that tropical trees in the average have higher ash content and therefore lower actual combustion values per unit dry-matter produced. A detailed analysis of both maps requires further equalization in the production of both maps, which will be done when new data become finally available from all stations of the International Biological Program.

ENERGY DEPRECIATION DURING DECOMPOSITION OF LITTER FROM DIFFERENT ECOSYSTEMS

All primary produced material will eventually die and drop to the bottom of the ecosystem where it is subject to decomposition. One of the more obvious characteristics in several ecosystems is the large amount of undecomposed dead material lying on the ground (e.g. boreal coniferous forest, peat bogs). Even in not so obvious cases large differences occur among ecosystems in the speed of decomposition. With decomposition of organic material the ash portion is released as well, and this serves as new nutrient input for the regrowth of plants. An optimal speed and timing of decomposition would therefore be desirable. However, too little is known about the decomposition process carried out by fungi, bacteria and invertebrate animals to apply any scientifically based management technique for the optimization of nutrient release from the dead biomass. If one investigates the energy content of the dead litter over time, one gets the following results from different ecosystems: The energy value of litter dropped in the fall rises during the first weeks and then slowly falls. This is almost entirely due to the initial leaching pattern of ash minerals from the litter and later on the relative increase of ash material insoluble in water while the organic matter slowly diminishes with the activity of the microbes. This effect has been shown repeatedly by Bocock (9), Runge (10) and others (11) in forest litter and by Britton, Dodd, and

Weichert (in press, 12) in natural grassland litter. The reports by Britton _et al_. are, with permission of the authors, included in the lower portion of Figure 3. The time sequence depicted there shows the

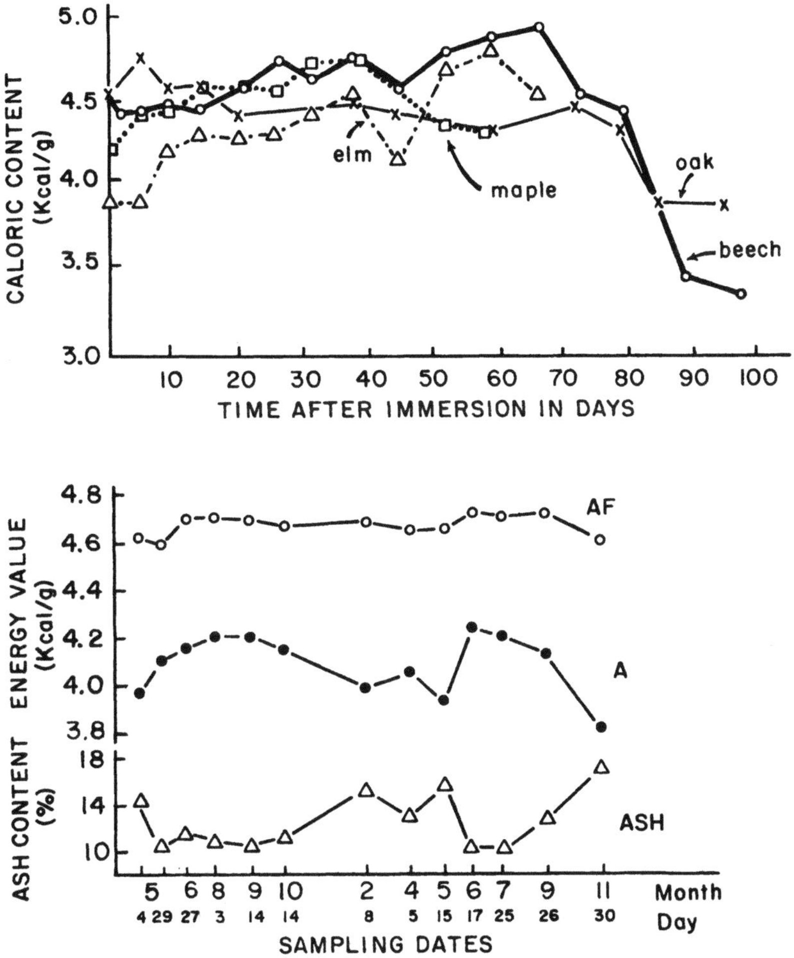

Figure 3: Changing combustion values of decaying litter in time. Upper figure: different broadleaf deciduous trees decaying under water, after Pena (1975), (15); abscissa - time in days; ordinate - combustion value of total dry-matter in Kcal/g. Oak - white oak, _Quercus alba_; beech - _Fagus grandifolia_; elm - American elm, _Ulmus americana_; maple - _Acer platanoides_. Lower figure: decaying litter of Andropogon-Paspalum grassland, after Britton _et al_. (12); abscissa - time in months and days; ordinates - combustion value for total dry matter (A), and ash-free organic matter (AF) in Kcal/g, and % ash for lowest curve.

insignificant fluctuation of the ash-free combustion values of grassland litter over 1.5 years and the almost perfect mirroring of actual combustion values (including ash in dry weight) and of the % of ash in the dry-matter over the same time period.

Over a longer time period the litter breaks down into humus compounds of unidentifiable species origin. Runge (10) reports from the Solling project in Germany the following sequence of energy values (Table II):

TABLE II: Average combustion values in Kcal/g during the conversion of litter to humus (after Runge 1973), (10) averages from measurements over 1 year.

	Litter	Transition 1	Transition 2	Humus
Fagus silvatica				
Beech ash free	5.10	5.10	5.20	5.50
Total dry weight	4.90	4.75	4.50	3.70
Ash %	3.60	6.60	12.60	32.30
Picea abies				
Norway spruce ash free	5.25		5.55	
Total dry weight value	4.95		4.35	
Ash %	7.00		20.50	

While the energy values of the ash-free organic component increase in the transition from litter to humus, the actual combustion values for total dry-weight decrease because of the increasing ash content.

Kovda (13) and Ryszkowski (14) estimate the time span for litter breakdown in temperate deciduous forests to be 3-4 years and in coniferous forest to be 10-17 years. The biochemical changes during this time span are still unclear. The increase in energy content of the decomposing litter suggests that the cellulose (4.2 Kcal/g) is decomposed earlier than the lignin (6.3 Kcal/g).

When the decomposing microbes become an important portion of the litter analyzed, the energy content values may change. Runge (10) reports values for the fungal mycelia in the same order of magnitude as in the original litter but lower than those of the ash-free organic matter component of the humus. The latter observation is substantiated by Pena (15) (see upper part of Figure 3), who observed the actual energy values for leaves of different broadleaf forest species decaying under water. The leaf material increased in energy content at the beginning but sharply dropped after 70 days under water. At the time of the energy-content decline, microbial cover was observed on the leaves. The ash content was not observed at this time, and the organic chemicals involved were not analyzed. Further investigations are needed to explain the changes in energy fluctuations in most cases, and the best results may be expected in the future from a combined quantitative study of biochemistry and combustion values.

ENERGY TRANSLOCATION IN INDIVIDUAL PLANTS DURING GROWTH AND DEVELOPMENT

The first product of photosynthesis, sugar, is used to construct the plant body, as well as some reserve material. The many plant chemicals have vastly different energy values ranging from about 11 Kcal/g for isoprene in rubbertrees to 0.67 Kcal/g for oxalic acid with a medium value for sugar and cellulose around 4.2 Kcal/g. During the growth of a plant, especially towards the end of the vegetation period, high-energy reserve material is often synthesized. Under such conditions one should assume that the ratio energy-value/dry-matter for the total plant increases. The development of dry-weight over time and total energy-sum over time plotted for the same plant should show the dry-weight increasing faster than the energy-sum during the initial growth phase and leveling off later.

That such a relation exists was proven in our laboratory in an experiment with growing sunflowers (Kreh, 16 and Pflanz, 17). As shown in Figure 4, the dry-weight of the sunflowers increased faster than the amount of energy accumulation did. After 75 days the weight was considerably higher than the energy accumulated if the end-point of both criteria were used as a reference ratio. From that time on, the sunflowers developed seeds, which contain a high percentage of oil. With the translocation of assimilates into the seeds the disparity between dry-weight development and energy accumulation slowly diminishes.

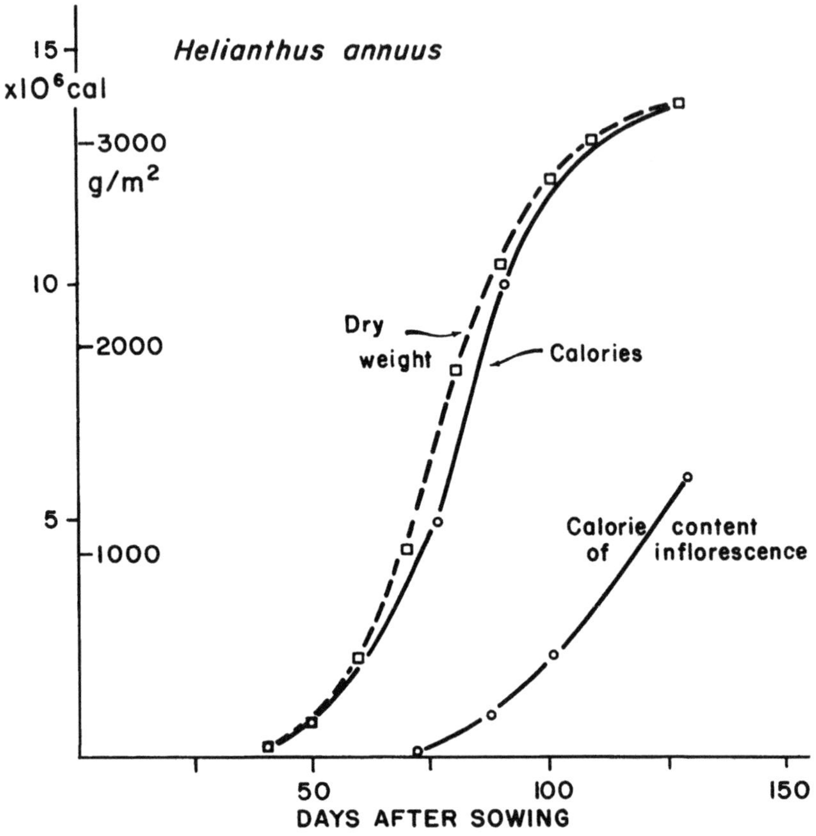

Figure 4: Dry-weight and total energy accumulation during the growth of sunflowers. Abscissa - time in days after sowing; ordinates - g/m^2 for dry-weight, dashed curve; 10^6 cal for calorie accumulation; the two ordinates are normalized for the highest measured value at 130 days after sowing. The comparison of the two curves shows the low-energy dry-matter at the early growth stage changing to higher energy levels when seeds are formed.

Such comparative analyses of energy translocation, accompanied by the biochemical analysis of the materials translocated, has great significance for studies of ecological energetics. In many cases the energy content of the food and the efficiency of assimilating the energy are the limitation for the growth of animal populations. It is for this reason that energy determinations have become increasingly important in ecology.

SUMMARY

The value of energy determination in ecological research has been demonstrated at three different organizational levels:

The global energy-fixation pattern is presented in the form of a table comparing energy fixation rates in different biome types (vegetation units). The rates reported were used to prepare a computer map showing the global pattern of energy fixation rates. Consistent with our previous mode this map is called the Berlin 1976 map (Table I and Figure 1).

The release of energy during the decomposition of litter in different ecosystems is discussed. Changes in energy values on a short time basis appear to be caused by fluctuations in ash content. Over longer periods of time changes of energy values have also been demonstrated for the ash-free organic matter. The chemistry causing such changes, however, is still mostly unknown (Figure 3 and Table II).

The energy accumulation in relation to the dry-matter development of growing sunflowers shows interesting trends. In the earlier growth phase, materials of lower energy values are photosynthesized and apparently stored until flower heads are formed and large portions of the stored low-energy material is translocated and converted to oil in the seeds (Figure 4).

LITERATURE

(1) Lieth, H.: Versuch einer kartographischen Darstellung der Produktivität der Pflanzendecke auf der Erde. Geographisches Taschenbuch 1964/65, 72-80. Wiesbaden, Max Steiner Verlag (1964).

(2) Cummins, K.W., Wuycheck, J.C.: Caloric equivalents for investigations in ecological energetics. Mitt. Int. Ver. Limnol. 18, 1-158 (1971).

(3) Lieth, H.: Measurement of caloric values. in Lieth, H., Whittaker, R.H. eds. Primary productivity of the biosphere p. 119-129. New York, Springer Verlag (1975).

(4) Lieth, H.: Über die Primärproduktion der Pflanzendecke der Erde. Z. Angew. Bot. 46, 1-37 (1972).

(5) Lieth, H.: Primary production: Terrestrial ecosystems. Human

Ecol. 1, 303-332 (1973).

(6) Lieth, H.: Primary production of the major vegetation units of the world, in Lieth, H., Whittaker, R.H. eds. Primary productivity of the biosphere p. 203-215. New York, Springer Verlag (1975).

(7) Lieth, H.: A computer model of the world vegetation. Memorias de Symposia del 1° Congreso Latinoamericano y 5° Mexicano de Botanica p. 451-458 + 7 maps. Mexico. Sociedad Botanica de Mexico (1972).

(8) Box, E.: Quantitative evaluation of global primary productivity models generated by computers. in Lieth, H., Whittaker, R.H. eds. Primary productivity of the biosphere p. 267-283. New York, Springer Verlag (1975).

(9) Bocock, K.L.: Changes in the amounts of dry matter, nitrogen, carbon, and energy in decomposing woodland leaf litter in relation to the activities of the soil fauna. J. Ecol. 52, 273-284 (1964).

(10) Runge, M.: Energieumsätze in den Biocönosen terrestrischer Ökosysteme. Scripta Geobotanica 4, 77 p. Göttingen, E. Goltze Verlag (1973).

(11) Bandhu, D.: A study of the productive structure of tropical dry deciduous forest at Varanasi. 111 p. Ph.D. Thesis. Banaras Hindu Univ. Varanasi, India (1971).

(12) Britton, C.M., Dodd, J.D., and Weichert, A.T.: Energy values of plant species and litter of an Andropogon-Paspalum grassland. Journal of Biogeography (in press).

(13) Kovda, V.A. ed.: Biosfera i je ressursy. Moskva, Nauka (1971).

(14) Ryszkowski, L.: Energy and matter economy of ecosystems, in van Dobben, W.H. and Lowe-McConnell, R.H. eds. Unifying concepts in ecology, 109-126, The Hague, Dr. W. Junk (1975).

(15) Pena, A.I.: Caloric content changes of decomposing leaves in stream water. 22 p. M.S. Thesis. Bucknell University, Lewisburg, Penn. (1975).

(16) Kreh, R.: Untersuchungen über den Aufbau und die Stoffproduktion eines Sonnenblumenbestandes. 71 p. + tables. Diplomthesis T.H. Stuttgart (1965).

(17) Pflanz, B.: Der Energiegehalt und die ökologische Energieausbeute verschiedener Pflanzen und Pflanzenbestände. 42 p. Diplomthesis T.H. Stuttgart (1964).

Index

The pages cited are the first pages of the paper in which the key word appears.

Walter de Gruyter
Berlin · New York

Curtius-Roth (Editors)

Clinical Biochemistry

Principles and Methods

2 Volumes. 1974. LXIX, 1677 pages. With 177 tables, 362 figures, 3 colored plates and 4463 references. Bound DM 460,– ISBN 3 11 001622 2

Hilde Götz

Immunologische Plasmaprotein-Diagnostik

1973. Octavo. XII, 240 pages. With 106 illustrations and 6 tables. Bound DM 96,– ISBN 3 11 003772 6

Hilde Götz

Antigenität von Tumorproteinen

1972. Octavo. XII, 126 pages. With 19 illustrations and 6 plates. Bound DM 60,– ISBN 3 11 004052 2

H. Rainer Maurer

Disc Electrophoresis and Related Techniques of Polyacrylamide Gel Electrophoresis

2nd revised and enlarged edition 1971

Large-octavo. XVI, 222 pages. With 88 figures and 948 references. Bound DM 68,– ISBN 3 11 003495 6

(Working Methods in Modern Science)

Prices are subject to change

Mikrokalorimeter CALVET

Dieses **SETARAM-Mikrokalorimeter** mißt mit einer sehr hohen Genauigkeit die umgewandelte Wärmemenge zwischen dem zu untersuchenden System und seiner Umgebung.

Hierzu werden weit über 1.000 Thermoelemente eingesetzt, die ein exaktes Messen im μW-Bereich gestatten.

Charakteristiken der einzelnen Gerätetypen.

Standardmodell MS in vier Ausführungen:
2- oder 4-zellig à 15 ccm Nutzvolumen.
2- oder 4-zellig à 100 ccm Nutzvolumen.
Meßschwellendetektion: 0,1 μW oder 50 μJ.
Arbeitstemperaturbereich: Raumtemperatur bis + 200° C.

Tieftemperaturmodell BT-215
2-zellig à 15 ccm Nutzvolumen.
Meßschwellendetektion: 0,2 μW oder 200 μJ.
Arbeitstemperaturbereich: – 206° C bis + 200° C.

Hochtemperaturmodell HT-1000.
2-zellig à 15 ccm Nutzvolumen.
Meßschwellendetektion: 5 μW oder 2,5 mJ.
Arbeitstemperaturbereich: Raumtemperatur bis 1.000° C.

Ihr gemeinsames, **elektronisches Zubehör** in 19"-Einschubtechnik:

- Eich- und Abgleichgerät für JOULE und PELTIER-Effekt einschl. der Abgleichzellen, Modell EJP-30.
- Numer. Nanovoltmeter mit BCD – TTL – LOGIC – Ausgängen, für einen eventuellen Rechneranschluß, Modell NV-724-N.
- Temperaturregler und -Programmierer, Modell PRT-4000.
- Integrator mit Ausdruckeinheit.
- Schreibende Registriergeräte.
- 19"-NORM-Meßschrank zur Aufnahme der Einschübe und Registriergeräte, ausgerüstet mit den elektrischen Absicherungen.

SETARAM
101-103, Rue de Sèze, F 69006 LYON/France

Technisches Büro Deutschland
Berliner Ring 134
D-6072 DREIEICH

Walter de Gruyter
Berlin · New York

Dietmar Richter
(Editor)

Lipmann Symposium
Energy, Regulation and Biosynthesis in Molecular Biology

1975. Large-octavo. XX + 701 pages. With frontispiece, figures, tables, photographs and references.
Bound DM 150,– ISBN 3 11 004976 7

Horst Sund
Gideon Blauer
(Editors)

Protein - Ligand Interactions

Proceedings of a Symposium held at the University of Konstanz, West Germany, September 2–6, 1974

1975. Large-octavo. XIV, 486 pages. With figures, tables and references. Bound DM 145,– ISBN 3 11 004881 7

H. Fritz
H. Tschesche
(Editors)

Proceedings of the International Research Conference on Proteinase Inhibitors

Munich, November 4–6, 1970

1971. Large-octavo. XII, 304 pages. With 256 figures and 95 tables. Soft cover DM 108,– ISBN 3 11 003776 6

Prices are subject to change

Mikrokalorimeter-System LKB 2107

mit universellen Einsatzmöglichkeiten in der Mikrobiologie, Biochemie und Biophysik

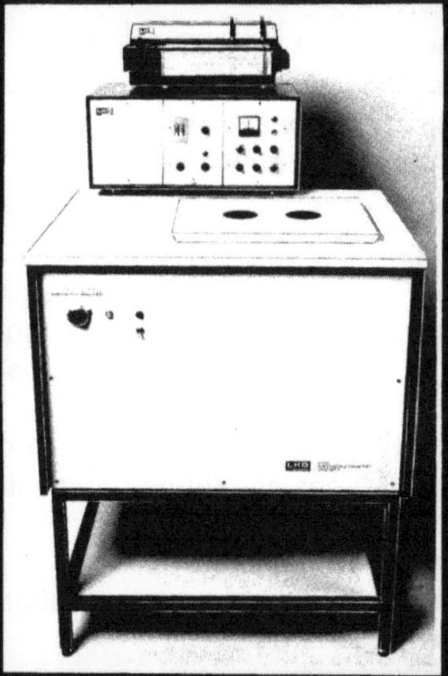

- Hohe Empfindlichkeit auf Reaktionswärmen — nur geringe Mengen an Probematerial erforderlich
- Isotherm arbeitendes Mikrokalorimeter — schnelle Einstellung des Temperaturgleichgewichts bei Änderung der Arbeitstemperatur
- Modulbauweise — je nach Aufgabengebiet können vier unterschiedliche Detektoreinheiten im Thermostatenbad mit elektronischer Kontrolleinheit verwendet werden
- Doppeldetektorsystem mit Kalibrierungseinheiten — direkte Kompensation z. B. von Reibungswärmen bei leichter Bestimmung der Reaktionsenthalpien

LKB Instrument GmbH
8032 Gräfelfing Lochhamer Schlag 5
Tel. 089/85 50 51 Telex 5 29 852
Berlin 030/7 95 85 53 Hamburg 040/2 80 30 40
Hannover 0511/80 51 80 Düsseldorf 0211/57 40 35
Frankfurt 0611/61 20 62 Karlsruhe 0721/84 11 91
München 089/15 39 84/85

LKB

Fordern Sie bitte weitere Unterlagen zum LKB 2107 Mikrokalorimeter-System an

W
DE
G

Walter de Gruyter
Berlin · New York

R. C. Allen
H. R. Maurer
(Editors)

Electrophoresis and Isoelectric Focusing in Polyacrylamide Gel

Advances of Methods and Theories, Biochemical and Clinical Applications

1974. Large-octavo. 316 pages. With 115 figures and 19 tables. Bound DM 105,– ISBN 3 11 004344 0

This book presents the most recent advances of the methods of electrophoresis (PAGE) and isoelectric focusing (PAGIF) in polyacrylamide gel which have gained wide use in all fields of biology and medicine.

The various chapters in this volume are comprised of papers presented at a conference held at Tübingen, Germany, October 6–7, 1972. This conference was arranged to bring together a group of specialists in the field in order to assess the state of the art, to discuss problems of standardization and optimization of separations, to relate theoretical considerations to practical application, as well as to discuss the limitations and future potential of these techniques in biology and medicine.
This volume should serve as a useful reference to investigators and students who wish to, or who are already employing PAGE and PAGIF in their work.

Prices are subject to change
For USA and Canada: Please send all orders to:
Walter de Gruyter Inc., 162 Fifth Avenue, New York, N. Y. 100 10

Bei Fragen zur Produktsicherheit wenden Sie sich bitte an:
If you have any questions regarding product safety,
please contact:

Walter de Gruyter GmbH
Genthiner Straße 13
10785 Berlin
productsafety@degruyterbrill.com